P9-DYY-855

GENESIS

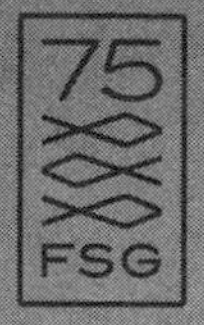

GENESIS

THE STORY OF HOW EVERYTHING BEGAN

GUIDO TONELLI

TRANSLATED FROM THE ITALIAN BY
ERICA SEGRE AND SIMON CARNELL

Farrar, Straus and Giroux
New York

Farrar, Straus and Giroux
120 Broadway, New York 10271

Printed in the United States of America
Originally published in Italian in 2019 by Giangiacomo
Feltrinelli Editore, Milano, Italy, as *Genesi*
English translation originally published in 2021 by
Profile Books Ltd, Great Britain
Published in the United States by Farrar, Straus and Giroux
First American edition, 2021

Library of Congress Cataloging-in-Publication Data
Names: Tonelli, Guido, 1950– author.
Title: Genesis : the story of how everything began / Guido Tonelli ; translated from the Italian by Erica Segre and Simon Carnell.
Other titles: Genesi. English
Description: First American edition. | New York : Farrar, Straus and Giroux, 2021. | Originally published in Italian in 2019 by Giangiacomo Feltrinelli Editore, Milano, Italy, as Genesi. | Summary: "A breakout bestseller in Italy, now available for American readers for the first time, *Genesis: The Story of How Everything Began* is a short, humanistic tour of the origins of the universe, Earth, and life—drawing on the latest discoveries in physics to explain the seven most significant moments in the creation of the cosmos" —Provided by publisher.
Identifiers: LCCN 2020050642 | ISBN 9780374600488 (hardcover)
Subjects: LCSH: Cosmology. | Life—Origin.
Classification: LCC QB981 .T6813 2021 | DDC 523.1—dc23
LC record available at https://lccn.loc.gov/2020050642

www.fsgbooks.com
www.twitter.com/fsgbooks • www.facebook.com/fsgbooks

1 3 5 7 9 10 8 6 4 2

For little Jacopo

We need poetry, desperately.

Anonymous (graffitied on a wall in an
alleyway in central Palermo)

All sorrows can be borne if you put them into a story, or tell a story about them.

Isak Dinesen

To be rooted is perhaps the most important and least recognised of the needs of the human soul.

Simone Weil

Contents

GENESIS

Introduction: The Grand Narrative of Origins

Forty thousand years ago, when the second wave of *Homo sapiens* arrived from Africa, many areas of Europe were already populated by Neanderthals. Organised into small clans, they lived in caves that today provide unequivocal proof of a complex symbolic universe: walls painted with symbols and drawings of animals, bodies buried in foetal positions, bones and large stalactites arranged in ritual circles; plentiful evidence, in other words, of a culture that had in all probability also developed a sophisticated spoken language.

It is possible, then, to imagine hearing a story of the origins of the world already resounding in those caves, the elders transmitting it to the young through the power of words and the magic of memory: the echo of an ancient narrative. It would be thousands of generations before Hesiod (or whoever we know by that name) left us, in his *Theogony*, a written account of how our universe came into being, woven from poetry and cosmology.

This ancient origin story continues to evolve to this day, thanks now to the language of science. Equations might lack the evocative power of poetry, but the concepts of modern cosmology – a universe that was born from a fluctuation in the quantum void, or from cosmic inflation – can still take our breath away.

All such stories spring from one simple, inescapable question: 'Where does *all this* come from?'

It's a question that still resonates in every part of the world, among people belonging to vastly different cultures. It's a point of commonality in otherwise distant civilisations; a question posed alike by children and executives, scientists and shamans, astronauts and those small, isolated populations of hunter-gatherers who survive in areas of Borneo, Africa and the Amazon. A question so primal that some have even imagined it must have been handed down to us by the species that came before.

Foundation Myths and Science

For the Kuba of the Congo, the universe was created by the great Mbombo, the lord of a dark world who vomited up the Sun, Moon and stars in order to free himself from a terrible stomach ache. According to the Fulani of the African Sahel, it was the hero Doondari who transformed an enormous drop of milk into earth, water, iron and fire. For the Pygmies of the forests of equatorial Africa, everything came into being when a huge turtle laid its eggs while swimming in the primordial water.

At the origin of mythological narratives of this kind, there always seems to be something indistinct that most troubles us: chaos, darkness, a liquid and formless expanse, a tremendous fog, a desolate Earth – until a supernatural being intervenes to shape things and bring order. This is where the great reptile comes in, the primeval egg, the hero or the creator who separates the heavens from the Earth, the Sun from the Moon, and gives life to animals and men.

The creation of order is necessary because it establishes the rules, providing the foundations of rhythms that mark the life of communities: the cycles of day and night, the changing seasons. Primitive chaos triggers an ancestral fear: the terror of falling prey to the forces unleashed by nature – from ferocious beasts and earthquakes to droughts and floods. But once nature is induced to follow the rules dictated by the hero who has brought order to the world, humankind is able to survive and reproduce. The natural order is reflected in the social one, in the combination of norms and taboos that define what is allowed and what is strictly forbidden. If the tribe behaves according to the rules established by that originary pact, then this stockade of rules will protect the community and prevent its disintegration.

From such myths as this, other constructions follow. They morph into religion and philosophy, art and science – disciplines that will hybridise and fertilise each other, and allow enduring civilisations to flower. Yet their interweaving breaks down when science starts to develop out of all proportion to our other speculative activities. From that moment on, the sleepy rhythm of societies, unchanged for centuries, is ruptured by a succession of discoveries that will profoundly alter the existence of untold populations. Suddenly everything changes, and continues to change, at a vertiginous pace.

The emergence of science ushers in modernity. Societies become dynamic, subject to continuous transformations; social groups enter a period of ferment, dominant classes undergo profound changes, and in the course of just a few decades the centuries-old balance of power is disrupted.

But the most profound transformations do not concern

the way in which we communicate or produce wealth, our medicine or mobility. The most radical alterations are in how we look at the world, and therefore our place within it. The origin story provided by modern science very quickly acquires an unrivalled consistency and complexity. No other discipline can provide explanations that are more convincing, verifiable, coherent or consistent with regard to the myriad observations supplied by scientists.

And yet, despite the notion that we have progressively lost something of the magic and mystery that had accompanied us for millennia, the vision of the world that we have gradually developed through science is actually more amazing than anything we could have imagined before. It retells the story of our origins more imaginatively and powerfully than any mythological narrative. In order to construct this story, scientists have had to scrutinise the most hidden and minute corners of reality, and have explored the remotest worlds, coming to terms with states of matter so different from anything previously known to us as to nearly blow our minds.

From science we derive the paradigm shifts that define our epochs, and that irreversibly modify our relationships. And it is the ceaseless pressure of scientific discoveries that sets the tempo of this subterranean development, like the forces exerted upon the Earth's crust by white-hot magma, sometimes breaking it apart and irreparably transforming the landscape above.

Our lives are conditioned by the story of the origins of the universe told by science: it profoundly shifts the foundations on which we will build new social arrangements, opening up vistas full of opportunities and risks, and shaping the future for coming generations.

This is why, just as in ancient Greece everyone had to know the foundation myths of the polis, the origin story that is provided by science ought to be familiar to everyone. But for this to happen, we must first overcome a considerable obstacle. We need to come to terms with the difficulty of scientific language.

A Complicated Language

It all begins with an apparently marginal episode that took place just over four hundred years ago, and which has as its protagonist a Pisan professor of geometry and mechanics at the University of Padua. When Galileo Galilei began to modify the strange tube manufactured by a Dutch optician, converting it into an instrument for examining celestial bodies, he could hardly have imagined the trouble it would get him into; much less could he have foreseen the turbulence that his observations would create across the entire planet.

What Galileo sees through his system of lenses astonishes him. The Moon is not, after all, the perfect celestial entity described by the most authoritative texts. It is not composed of incorruptible matter, but has mountains, craters with jagged edges, and plains similar to our own. The Sun has stains on its surface and rotates on its axis; the Milky Way is a vast accumulation of individual stars; the 'starlets' surrounding Jupiter are orbiting satellites that resemble the Moon.

When in 1610 he publishes all this in his *Sidereus Nuncius*, he provokes, perhaps unwittingly, an avalanche that will undermine the system of beliefs and values that had prevailed for more than a thousand years and that no one had ever dared to challenge.

Modernity begins with Galileo: humankind frees itself from all protection and faces the vastness of the universe alone, armed only with its own ingenuity. A scientist no longer looks for truth in books, no longer bows his head before the principle of authority, no longer repeats formulas handed down by tradition – but subjects everything to the fiercest scrutiny instead. Science becomes creative research into 'provisional truths', through 'sensory experiences' and 'necessary demonstrations'.

The power of the scientific method resides in conjectures verified by means of instruments that allow the observation, measurement and categorisation of the most diverse natural phenomena. It is the results of these experiments, what Galileo calls 'sensory experiences', that determine whether a conjecture works or should be discarded.

From his observations, irrefutable evidence is soon provided in support of the 'lunatic' theories of Copernicus and Kepler, and our view of the world changes so radically that nothing will ever seem the same again. Art, ethics, religion, philosophy, politics – everything will emerge transformed by this conceptual revolution which places humankind and its capacity for reasoning at the centre of everything. The disorientations that this new approach will produce, in a relatively brief period of time, are so profound that it is hard to think of anything remotely comparable.

Galilean science is truly revolutionary because it does not allow itself the right to hold back the truth, but instead relentlessly seeks the possibility of falsification even in its own predictions; it welcomes the prospect of all the certainties established up to that point crumbling if they turn out to be false, and it is self-correcting in the light of experimental verification. Finally, in order to stress-test the increasingly complex conjectures that are being elaborated in its name, it pushes towards the exploration of ever more mysterious aspects of matter and of the universe.

From this patient and self-critical approach, accounts are produced of elusive and apparently marginal phenomena. In the process of evolving a view of the world that is ever more sophisticated and complete, we end up mastering, down to the minutest detail, the most remote natural phenomena – and at the same time developing increasingly sophisticated new technologies.

The price to be paid for this course of development is the need to use increasingly complex instruments, and a language increasingly remote from common speech. No sooner have we departed from the realm of daily life than the instruments and conceptual apparatus that govern our ordinary activities become inadequate. When we explore the minuscule dimensions in which the secrets of matter tend to be hidden, or embark on an exploration of those cosmic spaces that tell of the origins of the universe, we find ourselves in need of very special equipment and years of preparation.

This should come as no surprise. Even actual journeys of exploration and adventure across the globe require a great deal of preparation, effort, and specialised equipment. Think of extreme sailing, or climbing in the Himalayas, or descending into the oceanic abyss. Why should scientific exploration be any easier?

Anyone wishing to thoroughly appreciate physics, therefore, will need to labour for years, to study group theory and differential calculus, and gain a command of relativity and quantum mechanics, as well as field theory. These are all abstruse subjects, involving concepts and language that are difficult even for those who have used them for years. But the barrier of specialised language which prevents most people from entering into the living heart of modern scientific research can also be surmounted more readily. We can still use ordinary language to explain the basic concepts, and to make the vision of the world that science is in the process of advancing accessible to everyone.

A Dangerous Voyage

In order to understand the origins of our universe, we must be prepared to undertake a risky journey. The danger comes from the fact that we need to push our minds into areas or environments so remote from those we are accustomed to that our usual conceptual categories are no longer of any use. We find ourselves obliged to say the unsayable, to depict the unimaginable, to experience the limits of our mind. Limits of a mind that for *Homo sapiens* has been a powerful tool for exploring and colonising our planet, but which turns out to be altogether inadequate for fully understanding what happens in such vastly distant places. Like the explorers of old, we have no other option but to point the prow of the ship towards a horizon, and to accept the risks and the unknowns that go with navigating in uncharted waters.

Similarly, in scientific research the voyage home and return to port is also very important. In this respect the modern researcher is a lot like Ulysses, for wherever his journey takes him he is always dreaming of the moment when he will reach Ithaca again. Coming home also means, even if no new territory has been discovered, or we have suffered a terrible shipwreck, that it is possible to warn other sailors about the routes not worth taking, and the dangerous passages that should be avoided.

We do this because modern science is also a great collective adventure. We have theories and charts to guide us, but chance often takes us to places that are completely unknown. We have 'ships' that are meticulously cared for, but we only need to neglect one small detail and disaster can befall us. Our crew is a colourful and lively community of thousands of passionate minds, patient and curious modern explorers, quick like Ulysses to invent new stratagems to overcome whatever unexpected events might be thrown at them.

Despite the objectives of our research raising almost philosophical questions (What is matter made of? How did the universe come into being? How will the world end?), the practice of experimental physics is one of the most concrete activities imaginable.

A particle physicist – one of the thousands of researchers in the world exploring the behaviour of the extremely small components of matter – does not spend his time sitting at a desk making calculations, meditating on theory and fantasising about new particles. A modern apparatus for high-energy physics is as tall as a five-storey building, weighs as much as a cruise ship, and contains tens of millions of detectors. To construct and make operable these miracles of technology, thousands of people are required, and painstaking, obsessively detailed work that can take decades needs to be done. To devise new instruments more sophisticated than the previous ones, to prepare 'ships' more agile and swift for our navigation, we spend years producing prototypes, in relentless efforts to make them work before going on to build them on a vast scale. And even when detectors are rigorously cared for and installed in the experiment, functioning quietly for months, we are always faced with the fear of catastrophe. An overlooked minor detail, a defective chip, a fragile connection, a cooling tube that has been hastily soldered can at any moment cause irreparable damage to the entire collective enterprise. The difference between an outstanding scientific success and the worst of all possible failures frequently lies hidden in some stupid, insignificant detail or other.

The Two Paths of Knowledge

How do we collect experimental information on the birth of space-time? How do scientists study the first cries of the infant universe? Two paths of knowledge come into play here, completely independent and different from each other.

On the one hand there is particle physics, exploring the infinitely small. Its starting point is the matter that surrounds us – what rocks and planets, flowers and stars are made of, along with everything else besides, including ourselves. This matter has very special properties which may appear ordinary to us, but that are in reality very peculiar and in a way linked to the fact that the universe is a structure that is both very old and currently very cold. The most recent data tells us that our 'home' was built almost 14 billion years ago, and that we are talking about an extremely cold environment, reaching what seem like impossibly freezing temperatures. For us, isolated on the planet Earth, everything seems comfortably warm; but as soon as we leave the protective shell of our atmosphere, the thermometer plunges. If we measure the temperature at any point in the vast empty spaces that separate stars, or in intergalactic space, the thermometer registers just a few degrees above absolute zero, which is to say minus 270 degrees Celsius. The matter of the current universe – rarefied, extremely ancient and extremely cold – behaves in a very different way than when it was recently born and existed as an incandescent object of tremendously high density.

In order to understand what happened in those very first instants of life we need to be ingenious, to find a way of returning the elementary vestiges of current matter to the extremely high temperatures of those original conditions. We have to make a kind of journey back in time.

This is precisely what happens in particle accelerators. By making protons or electrons collide at very high energy, we exploit Einstein's equation: energy equals mass times the speed of light squared. The higher the energy of the collision, the higher the local temperature that will be obtained and the greater the mass of particles that we are able to produce and study. To reach the maximum energies possible we need truly gigantic equipment, such as the Large Hadron Collider at CERN that stretches for some 27 kilometres beneath the ground near Geneva.

Here we find that by heating extremely small portions of space to temperatures comparable to those of the primeval universe, extinct particles revive: those ultramassive particles that used to populate it and that vanished aeons ago. Thanks to the accelerators, they re-emerge for an instant from their icy tomb, as if from hibernation, and may be scrutinised in great detail. This is how we discovered the Higgs boson. We brought back to life several handfuls after they had slumbered for 13.8 billion years. Naturally, of course, the much sought-after bosons then immediately disintegrated into lighter particles, but they had left behind tell-tale traces in our detectors. The images of these special kinds of decay accumulated, and when the moment came when we were certain that the signal was well differentiated from the background, and that the other possible causes of error were under control, we announced the discovery to the world.

The exploration of the infinitesimally small, the reconstruction of extinct particles, the study of the exotic states of matter that everything was made up of at first; these constitute one of the two paths available for understanding the very first moment in the life of space-time. The other requires super-telescopes, huge instruments for exploring the infinitely *large*, and it studies stars, galaxies and clusters of galaxies, in an attempt to encompass the entire universe. Here too we resort to Einstein's equation in which the speed of light is fixed at approximately 300,000 kilometres per second: an extremely high but not infinite speed. Hence when we observe a very distant object, galaxies that are distant from us by billions of light years appear not as they are now (and it is quite difficult to define what 'now' means), but as they were billions of years ago, when they actually emitted the light that has only just reached us.

Looking with super-telescopes at very large and very distant objects, it is possible to watch all the principal phases of the formation of the universe 'live', and to collect valuable data about our history. In this way, by observing the first faint signals emitted from the heart of enormous gas clouds, we can understand how stars are born: we observe the thickening of gas and dust in the rings of material that orbit around some new celestial body, indicating protoplanetary systems in formation. This is how our Sun was born, and how the planets that orbit around it were formed – and it is truly amazing to see it actually happening.

Pushing a little further, we can witness the formation of the first galaxies, turbulent objects that sometimes emit enormous quantities of radiation of all wavelengths, an unequivocal sign of traumatic birth. Through the super-telescopes we can ultimately both observe the wonders of the universe and measure some of its properties with unbelievable accuracy. The local distribution of temperature throughout the universe is like a kind of incredible memory containing eloquent traces of what happened at the very beginning: extremely small fluctuations in temperature speak of our most remote history, in a language that we have managed with time to decode and interpret.

But the most awe-inspiring thing of all is that these two paths of knowledge, based on methods so different from each other that they are almost entirely distinct – undertaken and developed by two wholly independent scientific communities – are nevertheless completely coherent with each other: the data gathered from the world of infinitely small elementary particles, and that which pertains to enormous cosmic distances, converge implacably towards the same story of origins.

Abandon All Prejudice, Ye Who Enter Here

Scientific discourse requires, above all else, the abandonment of every kind of prejudice. Genuine explorers do not fear the unexpected. Far from it, they cannot wait to find themselves faced with utterly unforeseen phenomena. Like the Argonauts setting sail in search of the golden fleece, they are inspired more by curiosity than by the prospect of reward. They are not looking for safety, they are deliberately embracing risk.

When we undertake to journey towards the origin of the universe, as we are about to do, the concepts that guide our everyday lives, such as the persistence of things, the reassurance we feel when witnessing the harmony around us, must be left behind immediately and for good. We will no longer be able to refer to the universe with the word *cosmos*, as when everything seemed to belong to a well-regulated and orderly system that we were used to contrasting with *chaos*, the disorder relegated to remote and insignificant corners.

We are so conditioned by our everyday life, by what we see and experience habitually in the thin spherical shell we inhabit, that it is natural to imagine that the laws determining our existence are the same as those which prevail in every other corner of the universe. Spellbound by the regularity with which night follows day, by the recurrence of lunar cycles and the rhythm of the seasons, by the persistence of the stars that light up the heavens, we have assumed that a similar kind of balance must obtain everywhere. But this is far from being true. In fact, something like the opposite is the case.

We have only been here for a few million years, living lives of infinitesimal duration compared to the cycles of any relevant cosmic process. We live on a tepid, rocky planet, rich in water and surrounded and protected by an accommodating atmosphere and by a benevolent magnetic field which, like some kind of magic blanket, can absorb ultraviolet rays and screen us from the devastating impact of cosmic rays and swarms of particles. Our mother star, the Sun, is a medium-sized star located in a very calm and rather peripheral region of the galaxy which harbours us. The whole of the solar system orbits slowly, in a manner of speaking, at a distance of 26,000 light years from the centre of the Milky Way. A safe distance, because hidden within it there is a monstrous black hole, Sagittarius-A*, an object weighing 4 million times more than the Sun and capable of destroying thousands of stars in its vicinity.

If we now carefully observe further the phenomena that affect the celestial bodies, such as stars that seem to be stationary and inactive, we stumble across incredible objects and discover that immense quantities of matter may behave in a very strange manner.

This is the case with pulsars, dark and dense objects that concentrate in a radius of around 10 kilometres the mass of one or two Suns. Myriads of neutrons are held captive by the gravity of pulsars, which compresses and seeks to crush them while the star itself revolves in a vortex, producing tremendous magnetic fields.

Even worse than pulsars are quasars and blazars, ultra-massive bodies that roar at the centre of some galaxies. Disproportionately massive black holes – over a billion times more massive than our Sun – they are capable of swallowing the ill-fated stars that end up trapped in their gargantuan gravitational fields. This *danse macabre* has developed over millions of years, and we can observe it from Earth because the matter that is precipitated, spiralling, contorting into the abyss and disintegrating, ends up emitting high-energy jets and gamma rays that our detectors are able to identify.

These strange celestial bodies, neutron stars and black holes, are the cause of enormous catastrophes that occur regularly throughout the 'cosmos'. Today they can be studied with a great deal of precision, to the extent that we have even witnessed collisions between them that have distorted space-time, producing gravitational waves that have reached us from billions of light years away.

But we do not need to look as far as this in order to understand how beneath the appearance of the cosmos, there is a chaos concealed. We only need to look closely at the surface of the Sun. What seems to be a quiet star that calmly illuminates our days, when seen close up becomes a complex and chaotic system made of innumerable thermonuclear explosions, convection currents and periodic oscillations of awesome masses, with flows of plasma projected all around it by powerful magnetic fields. Within our mother star there is a clash occurring between titanic forces, a battle that has lasted for countless years, with only one winner: gravity. And in a few billion years, with the exhaustion of its nuclear fuel, it will finally succeed in crushing and shattering its own internal structure, causing it to collapse. The central nucleus will become compressed, while the external strata will begin to expand until they reach Mercury, Venus and Earth, causing them to instantly evaporate.

This is because systems that are markedly chaotic may appear orderly and uniform when seen from a great distance. The same thing happens at the other extreme of our observations, in the world of the infinitely small.

If we look closely at the most apparently smooth and polished surfaces, we are immediately struck by the chaotic dance of the elementary components of matter which fluctuate, oscillate, interact and change nature at a frenetic pace. The quarks and gluons that make up protons and neutrons are constantly changing condition, interacting among themselves and with millions of virtual particles around them. At a microscopic level, matter relentlessly follows the laws of quantum mechanics, dominated by chance and the principle of uncertainty. Nothing stays still. Everything seethes in an extraordinary, constantly changing variety of states and possibilities.

But when we observe great numbers of these particles, when the structures become macroscopic, the mechanisms that regulate this dynamic almost magically acquire regularity, persistence, order and equilibrium. The superposition of a sufficiently large number of random microscopic phenomena, developing in all possible directions, produces macroscopic states that are orderly and that endure.

Perhaps the time has come to introduce a new concept to describe this fact, which seems to be truly structural. *Cosmic chaos* might be the right oxymoron to capture the relation between the two entities that in the universe chase each other and play at hide-and-seek. It is a game that we observe when we probe the tiniest recesses in the world of elementary particles, but that also occurs when we observe what happens at the heart of stars or of gigantic structures such as galaxies or clusters of galaxies.

To understand the birth of the universe, we will need to abandon the prejudice of order, among many others. We will face a voyage guided only by imagination, and we will have resort to concepts so bold as to make the most fantastic science fiction seem banal by comparison. On this journey we will get to know about theories that are changing forever our view of the world. And at the end of it, perhaps, discover that we ourselves have become different from what we were at the beginning.

So fasten your seat belts. We are about to take off.

In the Beginning Was the Void

In the beginning was the void. There, we have done it, we have right away given an answer to the most difficult of questions: what was there before the Big Bang?

Strictly speaking it's a badly put question. As we shall soon see, space-time makes its entry together with mass-energy; hence there is no *before*, there is no clock that ticks *beyond* the universe that is still waiting to be born. Nevertheless, for the sake of the narrative we can ignore this logical difficulty and go straight to the substance of the matter.

Let us accept the paradox of asking ourselves what existed *before* time came into being and let us imagine ourselves in that *no-place* from which all space would unfurl. Let us fantasise, we material beings who need air to breathe and light to see, that we are already present there, where there is still no trace of matter or energy, waiting to assist at the birth of everything, and to see it with our own eyes.

Before us extends the void, a very peculiar physical system that despite its frankly misconceived name is anything but empty. The laws of physics fill it with virtual particles that appear and vanish in frenzied rhythm, packing it with fields of energy whose values fluctuate continuously around zero. Anyone can borrow energy from the great bank of the void and live an existence the more ephemeral the greater the debt acquired.

From this system, from these fluctuations, a material universe may emerge that in reality is still only a void, but a void that has undergone a marvellous metamorphosis.

A Gigantic Expanding Universe

Today it is hard not to smile a superior smile when confronted by the naive imaginings that the best scientists from previous eras relied upon, produced without the benefit of modern telescopes. The word 'universe' contains the Latin roots *unus*, 'one', and *versus*, the past participle of *vertere*, 'revolving'. We use it as a synonym for 'everything', even if its literal meaning should be 'that which is turned wholly in the same direction', containing therefore a residue of all the ancient beliefs that invariably involve a stable and ordered system of rotating bodies. This prejudice is shared by both the ancient conceptions of Aristotle and Ptolemy and the more modern models of Copernicus and Kepler.

From a conceptual point of view, the geocentric and heliocentric universes are completely different. For almost two thousand years' some of the world's most learned men engaged in calculations and endless disputes concerning the movements of the wonderful concentric spheres in which the Moon, Sun, planets and stars were fixed. Then, suddenly, this vision of the world collapsed.

Removing the Earth from the centre of creation was no trivial matter. For sixteenth-century society it brought with it a terrible cultural, philosophical and religious shock. From that moment onwards the world would never be the same again. Nevertheless, if we look at things from a certain distance, the two systems which appear to be so irreconcilable that blood was spilt over the gulf between them, are actually in possession of similar structures. Both describe an immutable, stationary universe; a perfect machine that guarantees perennial, harmonious rotation. Whether it is made to work by 'the love that moves the Sun and stars', or by the gravitational force of Galileo and Newton, the basic substance of what is involved does not change.

This prejudice or fixed idea of an eternal and immutable universe that's perfect in every respect, identical to itself *ab initio*, from its very beginning, is a legacy that reaches down almost to our own time. It is astonishing to come across it at the start of the twentieth century, even in the first formulations of relativistic cosmology.

In 1917 Albert Einstein, developing the consequences of his general theory of relativity, postulated a homogeneous, static, spatially curved universe. Mass and energy warp space-time, and would tend to make it collapse into a point – but if you add to the equation a positive term that compensates for this tendency towards contraction, the system remains in equilibrium. The beginning of modern cosmology is ushered in with this manoeuvre. To avoid the catastrophic ending of the universe – the inevitable result if only gravity were present – an arbitrary term was invented. Wanting to maintain the prejudice regarding the stability and persistence of the universe that had lasted for millennia and still evidently held Einstein captive, he forcefully introduced the 'cosmological constant', a kind of vacuum energy which is positive and tends to push everything outwards, thus contrasting with and counterbalancing the gravitational pull and guaranteeing the stability of the whole.

Today, now we know that the universe is made up of 100

billion galaxies, it is shocking to realise that scientists in the first two decades of the last century, among them some of the most brilliant minds of all time, were still convinced that it consisted solely of the Milky Way. It was the slow concentric movement of the bodies belonging to this galaxy that gave the idea of a universe that was like a stationary, harmonious and ordered system. Soon afterwards this was brought into question by new kinds of observation, but a radical break with the old conceptions was also anticipated by the brilliant intuition of a young Belgian scientist.

In 1927 Georges Lemaître was a thirty-three-year-old Catholic priest with a degree in astronomy from the University of Cambridge, and in the process of completing his PhD at the Massachusetts Institute of Technology. He is among the first to grasp that Einstein's equations can also describe a dynamic universe, a system of constant mass but one that is expanding – with a radius, that is, which gets bigger with the passage of time. When he presents this idea to his older and much more established colleague, Einstein's response is shockingly negative: 'Your calculations are correct, but your physics is abominable.' So deeply rooted is the prejudice which for millennia had conceived of the universe as a stationary system that even the most elastic and imaginative mind of the period rejects the idea that it can be expanding, and that as a consequence of this expansion it must have had a beginning.

It would take years of discussion and fierce argument before this extraordinarily novel idea was generally accepted by scientists, and a great deal more time would have to pass before it became public knowledge.

The key to its success is suggested by Lemaître himself,

in an article in which he proposed his new theory, backed up with measurements of the radial speed of extra-galactic nebulae.

At the time, the attention of astronomers was concentrated on those peculiar objects resembling clouds which they conceived of as being groups of stars aggregated together with agglomerations of dust or gas. Today we know that they are in fact galaxies, each containing thousands of stars, but the telescopes of the time were not sufficiently developed to show them in much detail.

In order to calculate the speed at which a star or any other luminous body moved, astronomers had long known how to use the *Doppler effect.* The same phenomenon that we notice with sound waves from an ambulance siren can be observed with light waves. When the source recedes, the frequency of the waves that we receive is reduced: the sound of the siren gets fainter the further away it is. In the same way, the colour of visible light shifts towards red with distance. By analysing the spectrum of luminous frequencies emitted by various celestial bodies, we can measure for each one this shift towards red, precisely the so-called *red shift*, and work out from this the radial speed with which they are receding from us.

But it was not easy to measure how far away these formations were, or consequently to determine whether they were situated within our galaxy or not. The solution was discovered by Edwin Hubble, a young astronomer working at the Mount Wilson Observatory in California, equipped with what at the time was the world's most powerful telescope.

The technique employed was based on the study of Cepheids, pulsating stars of variable luminosity or brightness. Hubble begins his work just a few years after the death of Henrietta Swan Leavitt, one of the first American astronomers, a young scientist who had contributed enormously to this field and received, as is often the case, no appropriate recognition. In fact, at the beginning of the twentieth century it was considered unthinkable that a woman could use a telescope, and the extremely rare young female scientists were often deployed in subordinate roles. Leavitt was entrusted with the role of human 'computer', a wholly secondary and badly paid job: her task, in fact, consisted of examining, one after another, thousands of photographic plates containing images taken through telescopes, and recording the characteristics of stars and other celestial objects. She was assigned, in particular, the task of measuring and cataloguing the apparent brightness of these stars.

The young astronomer focused her studies on the stars with variable luminosity belonging to the Small Magellanic Cloud, a nebula which at the time was thought to be part of our own galaxy. It was Leavitt's incisive observation that the brightest stars were also those with the largest pulsation period. Once this correlation was established, an estimate of the absolute brightness of a star could be obtained, which in turn would allow us to measure its distance from us. The brightness of an object varies according to the inverse square of the distance from the observer, so by knowing its absolute brightness, one need only measure the apparent brightness to calculate the distance.

Leavitt measured the relation between luminosity and period in the Cepheid variables of the Small Magellanic Cloud, and by hypothesising that the stars were largely at the same distance, she was able to construct a scale of intrinsic luminosity, starting from the visible ones recorded on the plates.

Thanks to the incredible intuition of this brilliant young astronomer, we have at our disposal *standard candles*, that is to say light sources of known intensity, through which it is possible to deduce an absolute measure of distance.

This is what Hubble did when he used the Cepheids of the Andromeda nebula to reach the conclusion that these celestial bodies are too far away to be part of our Milky Way.

Lemaître was familiar with the first measurements made by Hubble, which not only placed these nebulae beyond our galaxy but also endowed them with an impressive speed of recession. His theory of an expanding universe made it possible to explain these unprecedented observations, as long as it was accepted that an enormous system was involved, immensely bigger than anything previously supposed. A gigantic structure in which there are countless galaxies similar to our own, with everything inclined to move away from everything else.

After having placed the Earth at the centre of the universe for thousands of years, and having reluctantly accepted that it is just one of the many bodies that rotate around the Sun, a further, final illusion suddenly crumbles. The solar system and our beloved Milky Way have no special position. We are an insignificant component of an anonymous galaxy – just one among the myriad of others to be found throughout the universe. And as if this was not enough, the entire system changes over time. Like all material objects it had a point of origin, and it will in all probability also have an end.

The Big Bang

Lemaître's intuition, confirmed by Hubble's measurements, provided the basis for nothing less than a new vision of the world. In his original article, written in French, the astronomer-priest had gone so far as to predict a relationship of strict proportionality between distance and the speed at which astronomical objects recede. If his idea about the expanding universe was right, the more distant galaxies would have to move away from us at higher speeds, and would consequently exhibit a greater red shift. And this is precisely the result that Hubble obtained as his catalogue of observations grew in complexity and richness. But for a long time, Lemaître's intuition was ignored because the Belgian journal in which he'd published his article had such a limited circulation. For this reason, until very recently, the scientific world had always referred to this correlation as 'Hubble's law'. Thanks to a careful work of reconstruction, the contribution of the Belgian scientist has finally been recognised. It took almost a century, but today the relation that made it possible to establish the essentially dynamic nature of the universe is called, appropriately, the 'Hubble–Lemaître law'.

In the early 1930s, faced with large amounts of experimental data, Einstein also ended up abandoning his initial scepticism. Legend would have us believe that when reluctantly admitting that the Belgian priest and the American astronomer were right, the eminent scientist regretted his previous failure to understand, remarking that the cosmological constant 'had proved to be the biggest blunder I have made in my life'.

Starting from an initial state in rapid expansion, there was no need to introduce this *ad hoc* correction, and so the cosmological constant disappeared for many decades from the fundamental equation of cosmology. By an irony of sorts, however, there would be a further reversal in the second half of the twentieth century when the discovery of dark energy caused the term that had so tormented its inventor to be reintroduced.

It was Lemaître once more who was the first to speculate that the expansion of the universe could actually be accelerating – and who, not by chance, kept Einstein's cosmological constant in the equation, albeit reduced to a much lower value. Lemaître described the birth of the universe as a process that had taken place some time between 10 and 20 billion years ago, starting from an elementary state which he called the 'primeval atom'. His hypothesis drew together the most advanced scientific theories of the period and the numerous mythological narratives that made everything originate from a kind of cosmic egg. But before doing so, he established the connection between microcosm and macrocosm that would prove so very fruitful in the coming decades.

From the outset, the formulation of this groundbreaking theory produced a great deal of perplexity. In truth, world opinion was otherwise engaged: the Wall Street Crash of 1929, the emergence of Fascism and Nazism in Europe, the many indications that the entire planet was about to descend into another global war. But even in scientific contexts where there was interest, the scepticism directed at the new cosmological theory was extremely strong. A good number of the most eminent scientists of the age refused even to countenance the idea of a *beginning* to space-time, or a *birth* of the universe. The problem lay in the fact that it bore a terrible resemblance to the biblical Genesis, and to the creation theory advocated by many religions. And if this wasn't bad enough, the first proponent of the theory happened to be a priest as well as a scientist, and a Roman Catholic one at that.

The idea of an eternal universe, of an uncreated and everlasting stationary state, had first been supported by Aristotle, and it still fascinated many scientists. One of the best known of these was Fred Hoyle, a British astronomer who simply considered the theory proposed by Lemaître to be utterly repugnant. Hoyle remained convinced by his own ideas right up until his death in 2001. In 1949, in a programme made for BBC radio, it was Hoyle who coined the term 'Big Bang' – a description he intended as derogatory. Ironically, the image of a great explosion that Hoyle had used with the intention of ridiculing the new cosmological theory ended up penetrating so deeply into the collective imagination that it contributed significantly to its success.

One of the bastions of the most tenacious opposition to the theory was provided by Soviet science. For decades, Soviet scientists stigmatised the Big Bang as a pseudoscientific and idealistic theory that hypothesised a form of creationism – far too similar to the religious kind not to be deeply suspect. It mattered little, for them, that Lemaître had scrupulously separated science from faith, to the extent of reacting with horror when in 1951 Pope Pius XII could not resist the temptation of referring to the Big Bang described by scientists as resembling the biblical moment of Creation. It was an attempt by the Pope to provide a sort of scientific basis for creationism, to reinforce the rational basis of faith – and Lemaître strongly objected to it.

It was experimental results, once again, that would determine the definitive success of Big Bang theory. Among the theoretical developments of the new cosmological theory there had been, in the 1950s, the prediction of a radiation diffused throughout the universe: fossil waves, the relics of a moment in which photons had irrevocably separated from matter and that continued to fluctuate around us. These were very weak electromagnetic waves, stretched for billions of years by the expansion of space-time, an attenuated energy that would have given to the interstellar void a typical temperature of a few degrees kelvin.

The stunning discovery that confirmed this was made almost by chance in 1964 by the American astronomers Arno Penzias and Robert Wilson. The pair had been working for weeks to mend an antenna they hoped to use for radioastronomical observations in the microwave region, but they had failed to eliminate an annoying signal that seemed to be coming from every direction at once. At first they had assumed that it was interference caused by a radio station transmitting in the vicinity of the laboratory; then they had thought it might be electromagnetic disturbance connected with various activities in nearby New York. After even checking that a pair of pigeons that had nested in the antenna – leaving a coating of whitish dielectric material, also known as pigeon poo – were not responsible, they stopped searching and published their results in a short letter. The discovery of cosmic microwave background (CMB) radiation emanating from all directions and the observation that the universe had a temperature of a few degrees kelvin, that is to say around –270 degrees Celsius, sealed the success of the already indisputable new theory. Penzias and Wilson had effectively recorded the echo of the Big Bang, the mother of all catastrophes, the primal event, the proof that everything had begun 13.8 billion years ago.

A Universe That Emerges from the Void

If truth be told, however, even during the years in which it enjoyed its greatest success, with the term entering into common parlance and Big Bang stories cropping up everywhere – from TV programmes to children's comics – doubts were still circulating among scientists.

Despite ever more accurate measurements of CMB adding increasingly convincing pieces to the puzzle, there was a big fundamental question left to answer. Basically, the traditional theory of the Big Bang concealed a major problem: if the universe is born from a point into which a tremendous amount of energy and mass are concentrated, producing an extremely dense and hot system that expands at a furious rate, what physical phenomenon had managed to concentrate everything into that point in the first place? It is, in a sense, the same question to which Italo Calvino alludes in *Cosmicomics*, in his short story 'Everything in a Single Point': 'Every point of each one of us coincided with each point of everyone else, in a unique point in which we all dwelt.' A similar idea had inspired Jorge Luis Borges, years before, in his sublime story 'The Aleph'. The story takes its title from the first letter of the Hebrew alphabet, alluding also to the primordial number which contains all other numbers, and involves a small, mysterious sphere in which one can view an entire universe.

Beneath the surface of the apparently established theory

lurked the huge question as to what mechanism could have brought everything to such an exceptional condition, to a dimensionless point with infinite density and curvature – what physicists call a *singularity*.

There seemed, in principle at least, to be a solution to hand that was both simple and intrinsically elegant. The same equations describing an expansion against the pull of gravity could be used to describe the reverse process: an unstoppable contraction that would lead inevitably to a tremendous implosion or Big Crunch.

Under certain conditions, the expansion of the universe can be slowed by the force of gravity acting upon matter and energy, until the expansion is completely nullified and a subsequent phase of contraction begins. In such a case, slowly but inexorably, there would be a concentration of galaxies into clusters and, in every corner of the universe an increase in both the density and the average temperature of matter. So everything would eventually lead to enormous concentrations of black holes, radiation and ionised matter that could only collapse catastrophically into a region of increasingly small dimensions, becoming virtually pointlike. And thus we have the singularity that will give rise to another Big Bang, from which a new universe will be born, and so on; links in an infinite chain of expansion and contraction, the squeeze of an immense accordion that builds its melodies through temporal cycles of tens of billions of years.

The idea of extending to the material universe this cycle of life, death and rebirth, without beginning or end, calls to mind certain concepts shared by many oriental philosophies. The universe itself would be subject to *Saṃsāra*, the wheel of existence that imprisons living beings in a series of countless reincarnations. This elegant and symmetrical solution appeared to have the merit of resolving with a light touch the apparent violation of the conservation of energy, answering the question as to how everything in the universe had been concentrated into a singularity.

It was an escape route that lay open for a few decades, but one which lost consistency as astronomers and astrophysicists achieved more accurate measurements both of the speed at which galaxies were receding and of cosmic background radiation, results which ushered in the birth of a new era of precision cosmology.

We have known for some time now that the stars tell their stories in a much richer and more eloquent language than had previously been imagined. The most powerful optical telescopes were soon flanked by huge dishes pointing in the direction of deepest space, gargantuan ears listening to radio signals coming from unknown stars or emitted from far-off galaxies. It was the birth of radioastronomy, and with it we were able to discover entire new families of material bodies – mysterious objects that send out characteristic radio signals and for which we choose exotic names such as *pulsars* and *quasars*. It would take decades of research in order to understand that behind some of the phenomena are new states of aggregation of matter: the force of gravity that roars out from the heart of the most massive celestial bodies fractures matter into its most minute components, producing the incredible density of neutron stars and of black holes.

The evidence that the cosmos is flooded with photons of every wavelength, from radio waves of tens of metres to the subatomic dimensions of the most powerful gamma rays, has prompted scientists to construct ever more sophisticated equipment, located both on the Earth and sent into orbit around it, capable of registering the whole spectrum of electromagnetic waves. Increasingly precise maps of the cosmos have been made – and of its innumerable sources of radiation at all frequencies. This impressive mass of data has made it possible to study the universe comprehensively, as a physical system that may be subjected to interrogation and provide answers to the typical questions that arise in such cases: What is the total energy that it possesses? And how much impulse, angular momentum, and total charge does it have?

As we gradually acquire ever more precise data and reduce errors in measurement, a picture with some very surprising aspects to it emerges. The data tells us that the expansion of the universe will not stop: nothing indicates that it will reverse its flight and head back towards the Big Crunch. The average density of the universe is not sufficient to reach that critical value at which gravity would become the dominant force. Therefore we have to abandon the otherwise almost irresistible seduction of a cyclical universe, and turn to confront again the fundamental problem of how the first singularity occurred.

It is at this point that another, even more elegant, solution presents itself. It turns out that the universe is extremely close to a condition of total homogeneity and isotropy. The incredible uniformity of cosmic background radiation tells us that the universe does not have an appreciable curvature; the angular distribution of this radiation tells us that space follows the laws of Euclidean geometry: a light ray that crosses a region of the universe undisturbed by mass and energy travels in a straight line. This is what is called a flat universe, with no curvature. And since the distribution of mass and energy in the universe is intrinsically linked to the curvature of space and to its geometry, according to the laws established by general relativity, we can suddenly reach the conclusion that a flat universe such as ours is a zero total energy system.

In other words, the positive energy derived from the mass and energy present in the universe is neutralised by the negative potential energy provided by the gravitational field. If we attempted to calculate the total energy of the universe as a system, we would have to start to transform into energy the mass of all the stars in our galaxy and multiply the result by 100 billion galaxies. Then we would need to add in dark energy and that which derives from dark matter, about which we will talk in more detail below. Finally, we would have to transform into energy all the forms of matter and radiation that wander throughout the universe: intergalactic gas and photons, neutrinos and cosmic rays, heading down until we reached gravitational waves. The result of this notional calculation would surely be an enormous positive number.

Then, armed with considerable patience, we would have to consider the negative contribution made to the total energy by the gravitational field. The force of attraction between two bodies, whether between the Earth and the Sun or two distant galaxies, produces a bound system, that is to say, the two bodies remain trapped in a system of negative potential energy. In order to liberate one of the two components we would need to supply positive energy, typically kinetic energy, accelerating one of the two bodies until it reaches escape velocity, that value which would make it possible to reach potentially infinite distances, thus breaking away once and for all from the gravitational pull of its partner. It is what happens when we want to launch a satellite from the Earth to explore the limits of the solar system.

Since gravity acts upon the whole distribution of mass and energy in the universe, the negative number derived from the totality of bound states is also gigantic.

Now all that remains is to measure the difference between these inordinately large numbers. And the result that we obtain is truly astonishing: it amounts to zero. In short, the total energy of the universe as a system is the same as if it were a void.

All of this can hardly be due to pure coincidence. Especially since something similar happens for the total charge of the universe, for its impulse, and for its angular momentum. Everything is strictly equal to zero. To summarise, the universe has zero energy, zero impulse, zero angular momentum and zero electrical charge: all characteristics that make it remarkably like a void. At this point the scientists must capitulate, since if it looks like a duck, walks like a duck, and quacks like a duck, then for them it *is* a duck.

The most sophisticated and complete observational data that we have gathered so far tells us in a consistent way that the mystery of the origin of the universe is hidden in the simplest hypothesis, which among other things resolves at a stroke the question that seemed to cause the Big Bang theory to waver. In a universe with a sum total of zero energy there is no need to resort to the strange mechanism that concentrates enormous quantities of energy and matter into the initial singularity, because within that point the energy is nil, and within the system that unfurls from it and that we call the universe there is also zero total energy. The physicist and cosmologist Alan Guth, one of the prime movers of this theory, defines it as the most appealing example of a 'free meal' furnished by the quantum vacuum.

That the whole universe comes from a void, or to put it better, that it is still now simply a void state that has undergone a metamorphosis, seems to be the most convincing hypothesis offered by modern cosmology, or at any rate the most consistent with the countless number of observations that have been collected to date.

The Void, or Nothingness?

But what is the void? Many equate the void with nothingness. But this is a serious mistake. Nothingness is a philosophical concept, an abstraction, that irreducible opposite of being that no one has succeeded in defining better than Parmenides: 'Being is, and can never not be; non-being is not, and can never be.'

Nothingness recalls ancestral fears such as the commonplace and recurring nightmare of falling into a bottomless well; the notion of vacuity is synonymous with the absence of value, as in a *vacant* soul, an *empty* argument. The association between the concepts of void and nothingness also comes from the inevitable closeness, for those brought up in Western cultures, between the cosmological theory of a universe that is born from a vacuum or the void and the Judaeo-Christian concept of creation of the world *ex nihilo*, from nothing. In reality, as we shall soon see, the concepts involved are almost opposites; the void as a physical system is in some respects the reverse of nothingness.

On the other hand, the concept of the void has many points of contact with the notion of zero. The term comes from the Latin *zephirum*, which appears for the first time in the West in 1202, in the writings of the great mathematician Leonardo Fibonacci. He translates into Latin in this way the Arabic *sifr*, which means precisely zero or empty, however much that Latin equivalent seems to recall the Greek myth of Zephyros, the gentle breeze that announces the coming of spring.

In Arabic the original meaning of the term indicating the number zero had been preserved, having been introduced from India where it was called *sunya*, which is to say *empty*. The same root may be traced again in the *Śūnyatā*, or 'doctrine of emptiness', the fundamental concept of Tibetan Buddhism according to which material bodies are in reality divested of a singular and independent existence. The Indians were the first to introduce the concept of zero-empty. The concept is resorted to for the first time in a work written in Sanskrit in 458. The title of this work is the *Lokavibhāga*, the literal meaning of which is 'The parts of the universe', and it is curious that it is a treatise on cosmology; almost as if establishing from the start a connection between the concept of the void and the birth of the universe.

This should come as no surprise, given the role that emptiness occupies in Indian cosmogony and in myths of creation. Shiva is the creator deity but also the destroyer of the universe. When he dances, the whole Earth trembles and the entire universe falls to pieces, burning under the pressure of the divine rhythm. Everything dissolves until it is concentrated into the *bindu*, the metaphysical point outside space and time, the coloured emblem of which is worn on the foreheads of many Hindu women. Then the dot itself slowly dissolves and everything is dispersed in the cosmic void. The cycle is reprised when Shiva decides to create a new universe and resumes his dance. Once again, the divine rhythm produces vibrations always wider than the void that ends up swelling spasmodically, giving birth to a new universe that takes its place in the infinite cycle of creations and destructions.

This familiarity of the Indians with the concept of emptiness or the void allows us to better understand why they were the ones who for the first time conferred upon zero the properties of a fully-fledged number, and inspired by the positional system already adopted by the Babylonians, determined its definitive glory.

This contrasted with the Greeks, for whom both zero and the infinite were terrible concepts which, defying logic, threatened the established order of things. The ideal of perfection, Parmenidean being, was represented as a sphere, always unchanging and equal to itself in space and in time – and above all, finite. The finite, for the Greeks, was synonymous with perfection, while the very idea of zero was anathema. How could *nothing* ever be *something*? It was no accident that zero could evoke primeval chaos: it is the number that, when multiplied by another number, rather than increasing it annihilates it altogether, dragging it into its abyss. Things don't get any better when we try to divide by zero, for in this case as well the result is a logical absurdity: the infinite, the limitless, the great unbounded formlessness. Like the void, the infinite was equally horrifying to the Greeks, inextricably linked to zero. The concepts that defied logic and troubled the minds of philosophers were deemed to be inappropriate, even dangerous: they could go so far as to spread panic and provoke social unrest.

It was for this reason that a kind of taboo was built around zero in the West, one that was then extended to the void as well. We need to free ourselves of this prejudice, which still conditions our way of thinking, in order to understand the mechanism by which the universe may emerge from the void.

The void that we are speaking about is not a philosophical concept, it is a particular *material* system, one in which matter and energy are null. It is a state of zero energy, but it is a physical system like all others that can be investigated, measured and characterised.

Over the years physicists have carried out countless experiments on the void. Sophisticated experimental equipment is used to study its strange properties, with the aim of understanding in detail how the state of the void influences some characteristic magnitudes of elementary particles. There are even some who imagine that they will discover in the void new phenomena that, once mastered, could lead to new technologies.

Just as for all physical systems, for the void the uncertainty principle has a role to play, as it regulates the behaviour of systems at a microscopic scale. The energy and proper time of a given system, including that of the void, cannot be simultaneously measured with precision at our leisure: the product of their uncertainties cannot fall below a certain minimum value. When we say that the void has zero energy, it means that by carrying out a large number of measurements we obtain zero as the average result of these calculations; the individual measurements give fluctuating values, both positive and negative, that are different from zero, distributed along a statistical curve with an average value of zero. The uncertainty principle tells us that the smaller the interval of time in which the measurement is made, the greater the energy fluctuations that would result.

In reality this characteristic has nothing to do with the disturbance of the system that occurs during the measurement, but is due rather to something deeper, linked to the behaviour of matter at microscopic levels. The state of the void has strictly zero energy if observed on a very long timescale – in theory an infinite one – but in very brief time spans it fluctuates like all things, crossing through all of its possible states, however improbable, including those characterised by energy levels significantly different from zero. In short, the uncertainty principle allows for the temporary formation in the vacuum of microscopic bubbles of energy, just as long as they then rapidly disappear. The lower the amount of energy involved, the longer these anomalous bubbles last.

Hence if we imagine the behaviour of the void at a microscopic level, we must not think of something boring, static, and always the same as itself. On the contrary, the fine web of the void seethes in a myriad of microscopic fluctuations. Those that involve more energy will immediately dissipate, but if the borrowed energy is zero they can last forever.

Things become further complicated if we consider the presence of matter and antimatter. The quantum vacuum fluctuations can take the form of the spontaneous generation of particle–antiparticle pairs. The void as a result may be seen as an inexhaustible deposit of matter and antimatter. We can take advantage of indeterminacy due to the uncertainty principle and extract an electron from it; if it is immediately returned to its proper place, no one will notice.

We only need to be quick enough for this to be possible. The operation amounts to removing an electron along with a positron. Here we need to pay close attention, because the rule regarding conservation of charge does not allow exceptions, and is much stricter than the rule regarding conservation of energy. I cannot extract a solitary electron as it would change the characteristics of the entire state of the void, which would remain positively charged. I must always make a positron, the positive electron, jump out as well, so that the system's total charge remains in equilibrium. Basically, we need to extract from the void the same quantity of matter and antimatter so that the void will not protest. However, the problem of the energy relating to the particle–antiparticle pair remains: the lower the mass of the pair, the greater the amount of free time they have available. This comes to an end with the uncertainty principle sounding the bell and the two 'pupils' returning to the discipline of their classes.

This mechanism is not a principle of physics occurring at some abstract level: it is a material process verified daily in particle accelerators. By bombarding the vacuum with the energy of colliding beams, new particles are produced, the more massive the higher the energy of the collision. In this way large quantities of particles can be extracted from the vacuum, and may be used for quite diverse purposes: from the radioactive isotopes used as markers in nuclear medicine to the Higgs bosons produced in the Large Hadron Collider.

The void is a living thing, a dynamic and constantly changing substance, full of potential, pregnant with (particle/antiparticle) opposites. It is not nothingness; it is on the contrary a system overflowing with unlimited quantities of matter and antimatter. In some respects it really does resemble the number zero, just as the Indian mathematicians thought. Far from being a non-number, zero contains the infinite set of positive and negative numbers, organised in symmetrical pairings of opposite sign with a null sum. The analogy could be extended to silence, understood as a superposition of all possible sounds that cancel each other out in phase opposition, or to darkness that can come into being due to destructive interference of light waves.

The hypothesis that everything might originate from a quantum vacuum fluctuation emerges naturally when we consider that, in our universe, the negative energy due to the gravitational field cancels exactly the positive energy associated with mass. A universe with such characteristics may come into being due to a simple fluctuation, and the laws of quantum mechanics tell us that it can last forever. The universe with zero total energy constitutes an important alternative to the traditional theory of the Big Bang, and renders superfluous the presence of an initial singularity.

The Void and Chaos

In a way the science of the twenty-first century takes us back to the reality of the story told by Hesiod in his *Theogony*, and to a magnificent and striking verse containing the origin of everything: 'Chaos was first of all.' An assertion perfectly in keeping with the story told by science, as long as it was not expressed using the most conventional and widespread interpretation of chaos, namely disorder – an undifferentiated agglomeration. We should restore instead the original meaning of the word, with its echo of the Greek *chaino*, 'open wide', *chasko*, open-mouthed, or *chasma*, a vortex. So it becomes a gaping black throat, a bottomless abyss, a dark gorge, the enormous void that's capable of swallowing and containing everything.

This original meaning of chaos was used for a long time. The association of the word with disorder was a much later development, due first to the work of Anaxagoras and then Plato. Chaos becomes with them the vessel confining formless matter that is awaiting being put into order by a superior principle. It will be Mind, or the Demiurge, that will give form to that rough and vile material, building the cosmos: that organised and perfect system which regulates and governs everything. Ever since the development of this new meaning, it is the one that has dominated – for millennia.

But original chaos, understood as the void, is anything but disorderly. There is no more strictly ordered, regulated and symmetrical system than the void. Everything belonging to it is strictly codified, every particle of matter goes hand in hand with its corresponding antiparticle, every fluctuation dutifully observes the constraints of the uncertainty principle, everything moves to a cadenced and well-tempered rhythm, a perfect choreography without improvisation or virtuosic excess.

But somehow this perfect mechanism is interrupted, something suddenly breaks in and takes over the scene, then initiates abruptly the process that will produce an expanding space-time and the mass and energy which bend it.

The extreme order that governs everything shatters in a fraction of a second, and the tiny quantum fluctuation inflates disproportionately, pushed by a process that we call *cosmic inflation*. Many details of this phenomenon still escape us, starting from the identity of the material particle, the inflaton, which having been extracted from the void by a purely random mechanism, gave rise to the wonderful sarabande or stately dance we will encounter in the next chapter.

Day One: An Irresistible Breath Produces the First Wonder

Everything happens in an instant. A moment earlier, that microscopic structure which bubbles restlessly, exactly like the others that surround it, appears to us to be altogether insignificant.

Looking around, it almost seems that what we are seeing is an extremely thin foam. The myriad minute fluctuations of which it consists call to mind the primeval liquid of myth: *aphros*, the word for foam in Greek, provided the name for Aphrodite, born from the blood and sperm of Uranus. Chronos, his son, sliced off his genitals with a scythe, to avenge his mother Gaia, and threw them into the sea, making the placid waters of Cyprus seethe with a miraculous event.

From quantum foam something even more astounding than the goddess of love and beauty will be born: an entire universe. But still no one can imagine what is about to occur. Only 10^{-35} seconds have elapsed since the moment it formed; such an insignificant amount of time that it is impossible to even conceive of it. Everyone expects that the microscopic bubble absorbing our attention will return to the ranks in a disciplined way, just like all the others. But the intervention of an unstoppable breath makes it grow out of all proportion. With one blow the infinitesimal object that was fluctuating in a calm and orderly way, following the strict ritual of the uncertainty principle, begins to inflate uncontrollably in paroxysms. The craziness that overwhelms it affects the surrounding void; drawing it inexorably in, dragging it into the same mechanism. Things have moved at such an incredible pace that in order to see exactly what has happened we would need a moviola. But no equipment could capture stills and show us the details of such a rapid metamorphosis as this.

Then, without warning, everything is restored to calm, and the strange thing that by now seems to have taken on an independent life of its own continues to expand, albeit at an extremely reduced rate.

We have just been present at the birth of a universe – our own. The first day is at an end, and a universe has come into being that already has everything it will need to evolve during the next 13.8 billion years. Yet only 10^{-32} seconds have passed.

A Peculiar Primordial Field

The universe begins, then, with a tiny fluctuation in the void that, while it is expanding, becomes filled with a strange substance that makes it swell disproportionately.

The first to propose the theory that disorientated modern cosmology was Alan Guth, a young physicist with a PhD from MIT who at the age of thirty-two was looking for a position in some prestigious university. He had been invited to conduct a seminar at Cornell, one of the best academic institutions in the USA, and it was there that in 1979 he presented his revolutionary idea.

As we have seen, the traditional theory of the Big Bang, although confirmed in general by observation, left too many problems unresolved. The first skeleton in the cupboard was the original *singularity* from which everything had begun. We could not understand by which mechanism it had been formed, given that the Big Crunch had been excluded. In the 1980s it was known that there was not enough matter in the universe to exceed a critical density, one that could have triggered the momentous implosion. It was therefore concluded that the flight of the galaxies would have slowly decelerated as an effect of gravity, but without giving way to catastrophic gravitational collapse. In short, we were lacking an explanation as to how the great Bang had taken place.

In an object of insignificant dimensions that may be produced by purely random mechanisms, the force that choreographs the dance is gravity, acting as a force of attraction. In order to expand and initiate the Big Bang we would need an extremely strong gravitational repulsion, an *antigravity*, something akin to the cosmological constant that Einstein had introduced in his equation to account for the stability of the universe – but something more awesomely powerful even than this.

Ordinary matter, mass and energy, produce a negative energy from the void; from which a positive pressure is born that tends in effect to squash and to encompass everything. If instead a completely new substance comes into play, responsible for producing positive energy, the pressure that derives from it is negative, which is to say it pushes outwards, tending to expand.

Another mystery is linked to the incredible homogeneity of the observable universe. Everywhere around us there are galaxies of every possible form; some that are calm and quiet while others are tormented by the pyrotechnics of *supernovas*, neutron stars and black holes. Yet however amazing this may seem, the cosmic landscape is actually quite repetitive. Generally speaking, when regions of great dimension are observed, we find that the objects contained in even the most remote corners of the universe are quite similar in kind.

It's a fact that calls to mind the sense of disorientation one feels when disembarking at airports located on other continents, in 'remote' places such as Kuala Lumpur or Sydney, only to find oneself walking among shops displaying in their windows exactly the same items of clothing we had left behind in Rome or Paris. The same goes for luggage, telephones and cameras. There is an obvious enough explanation for this, involving the great networks of distribution in a globalised world. On the other hand, until the 1990s no one had any idea about the mechanisms which had produced the incredible degree of homogeneity seen in astronomical observations.

The mystery seemed to further deepen because as ever more powerful telescopes were developed and used, making it possible to investigate areas of the universe that had until only recently remained inaccessible, the more we continued to see things that tended to look remarkably similar to everything we already knew: galaxies resembling those already seen, conglomerations of galaxies that seemed to be the twins of those just catalogued.

Even more astonishing was the uniformity of temperature measured in cosmic background radiation. Wherever the instruments were directed, the result never varied: 2.72 degrees kelvin, barely above absolute zero.

How was it possible that all the most remote corners of the universe, distant from each other by billions of light years, had agreed among themselves to attain exactly the same temperature at precisely the moment when scientists on a small planet in an anonymous solar system of an unremarkable galaxy had decided to take a look at what was happening around them? The distances between the regions of the universe being observed were far too great to speculate about any mechanism that could explain this phenomenon.

In order to find an answer, Guth tried to imagine what would happen if during the expansion of the elementary bubble, the resulting microscopic volume had been occupied by a positive vacuum energy, similar to what had been hypothesised for the cosmological constant. The most promising candidate for this seemed to be the Higgs boson, a particle that was being earnestly discussed at the time in order to explain the origin of the mass of elementary particles.

The Higgs is a neutral scalar particle with zero *spin*: unlike all the other elementary particles, it does not rotate on itself. In effect the Higgs field gives a positive energy to the void, but if the volume involved expanded at speed, the density of energy would decline as rapidly and could provide a kind of impulse. To maintain a constant density in a volume that is rapidly increasing, the total energy would need to grow in turn, and this would violate the principle of conservation of energy.

But what happens if during this steep decline there is an obstacle; if for some reason it stops for an instant on its race towards the zero potential, that of the void? Guth's response to this question once again changed the way we look at the origin of the universe.

An Unstoppable Expansion

The mechanism predicts a scalar field that gives to the void a potential positive energy, and during the course of its evolution it is blocked by a fraction of a second in a false vacuum state, a dip in the potential of constant and non-zero value.

Let's imagine an inexpert skier who goes down an easy piste slowly but must stop because he finds himself on a flat section or in a deep dip. For a while he will be trapped in the depression; he will have to push with his ski poles in order to emerge from it. Maybe he will fall and have to start over again before reaching the crest. Then, having overcome the slight negative incline, he will be able to resume his descent and quickly get to the bottom of the valley.

If the scalar field follows the example of the skier, that is to say it stops for an instant in the dip, a phenomenon of disproportionate violence is unleashed. Due to the positive vacuum energy, the bubble receives a push that will increase its volume. With the field blocked in the depression, the density of energy remains constant, and as the volume increases the positive energy that it had accumulated within it increases, and so ultimately does the impulse towards dilation.

Instead of removing it, the expansion movement emits energy into space. The bigger the bubble gets, the bigger the push towards expansion. This is the typical dynamic of exponential growth which, in this case, has a very convincing explanation. Thanks to excess energy, the bubble extracts from the void other scalar particles that fill its volume – and these, in their turn, augment the push.

Stuck in the dip, the field fills the space with a substance that exercises enormous pressure, not positive like that of matter and energy, but negative like the vacuum energy that Einstein had introduced with his cosmological constant.

For the great scientist it was enough to have a repulsive force that was relatively weak in order to counterbalance the force of attraction furnished by mass and energy, and it achieved the constant energy of the void: the field remained forever crystallised, like Snow White asleep in her glass coffin.

The primordial field hypothesised by Guth, on the other hand, has a strong dynamic; as in the fairy tale the kiss of the prince interrupts the beautiful girl's slumber – but it does so only for the briefest of instants, and unleashes a remarkable spell. That stealthy awakening which blocks the field of the false void for a fraction of a second produces a repulsive force that varies significantly over time. It assumes an immense scale in the period during which the field is blocked, and diminishes rapidly as soon as it exits from the false vacuum state. Alan Guth's anti-gravity, which causes an outbreak of furious expansion at the origin of the universe, is in the order of one hundred times the magnitude of the cosmological constant. It is this amazing negative pressure that has subjected everything to an expansion at a grotesquely accelerated speed. This is what the Big Bang originates from.

In an instant the unimaginable occurs. That minuscule object, billions of times smaller than a proton, undergoes an exponential growth which proceeds at such a ferocious pace as to make the most frenetic of Rossini *crescendos* seem ponderously slow by comparison. In the blink of an eye it becomes macroscopic. After emerging from this violent phase it is roughly the size of a football, and already contains all the matter and energy that it will need to evolve over the course of billions of years to come. In a ridiculously small fraction of time, this insignificant object has grown by tens of orders of magnitude, expanding at a speed much greater than that of light. The limits imposed by relativity – that nothing may move faster than the speed of light – are valid for whatever moves *within* space. For space itself, which expands in the void, or to be more precise transforms the void into space, these limitations do not apply. There is no speed limit for the newly born universe as it rushes towards its future.

Other quantum fluctuations, similar to those that had generated it, very soon free the nascent universe from the hollow in which it had been stuck, putting it on the right track and making it fall towards the true void state, which it reaches instantly. Since everything began, only 10^{-32} seconds have passed. But everything has changed utterly.

As soon as this phase has ended, while the field placidly oscillates in its hole of minimum potential, the energy that has been accumulating in the object that has undergone such an explosive transformation turns into a huge quantity of matter–antimatter, pairs of particles with their respective partners that are extracted from the void in great numbers and interact with each other and with the residue of the field until the whole reaches a state of thermal equilibrium.

The newly born universe now contains all current matter and energy, albeit concentrated into a small volume. Density and temperature are extremely high, and a second phase of expansion begins that, however rapid, proceeds at a decidedly less frantic pace than the one which prevailed only an instant before.

Alan Guth had effectively opened the oxhide bag that

Aeolus had gifted to Ulysses, containing the tempestuous winds that would impede his return to Ithaca. Like Ulysses' companions, he had removed the thin silver rope that kept it shut, thus liberating the most powerful of winds, releasing infernal forces.

In order to name this new phenomenon Guth resorted to the term 'cosmic inflation', from the Latin *inflare*, to inflate, which was already commonplace in economics to describe a steep rise in prices.

This better-known economic expression resonates with negative associations, derived from traumatic experiences of uncontrollably galloping price rises. We need only think of Germany in the aftermath of the First World War: the rise in prices was dragging everything into an uncontrollable spiral. Having just received their salary, workers had to rush to the market to buy everything that they could, because the next day their money would only buy half as much, and in a week's time their money would be reduced to so much worthless paper. The sellers, trapped in the same hellish system, had to continuously revise the prices of their goods to reflect rapidly increasing costs. In January 1923 you needed 250 marks to buy a kilo of bread; by December the price had risen astronomically to 400 billion marks. This is the potential absurdity of exponential growth.

The Success of Inflation Theory

The idea that the universe may have experienced a phase of cosmic inflation is still the subject of lively debate among scientists, even though a considerable majority believe it to be the most convincing explanation available.

One of the key strongpoints of the theory is that it provides an effective and natural explanation of the cosmological principle, in other words the extreme homogeneity of the universe at a macro scale.

At first glance the issue would seem to be quite counter-intuitive. We need only look up at the sky to see the Sun, Moon, planets and stars – giving us an immediate impression of the extreme variety of the structures that exist in the universe. In reality, however, this is one of the many prejudices of which we have become unwitting victims, simply due to the fact that we have a very confined point of view, and eyesight that does not allow us to see for vast distances.

But if we use the most modern exploratory instruments and expand our horizons until we can take in the entire cosmos, these 'local' differences become insignificant details. Recent experiments have catalogued 200,000 galaxies to conclude that, within a range of 100 million light years, the structures we encounter are always similar, to the point of being almost identical. Our universe, in short, though marvellously varied in its local contours, is somewhat monotonous, if not to say actually boring when navigated and looked at on a vast scale.

The homogeneity becomes even narrower if we consider temperature distribution. Since the 1970s the close study of cosmic background radiation has required programmes for using instruments carried by satellites. Freed from the disturbance caused by the Earth's atmosphere, it should have been possible to achieve much more precise measurements and above all to carry them out at all wavelengths. Yet it took another twenty years to obtain the first results that from the beginning of the 1990s furnished outstanding confirmation of the predictions made by the theory of cosmic inflation.

The homogeneity and isotropy of the universe are truly impressive. The temperature distribution replicates perfectly that which was predicted by the theory. The universe behaves like a giant microwave oven, the heat source of which ceased at some remote time in the past; since then it has cooled in a uniform way as it has gradually expanded. Regions separated by billions of light years have exactly the same temperature, measured with an absurd precision: 2.72548 degrees above absolute zero. The radiation is isotropic, that is to say it is the same in every direction, to better than one part in over a hundred thousand.

What mechanism made it possible for energy to be exchanged across such distant zones, to the extent of thermalising everything to such an incredibly uniform degree? It cannot be light alone, because when light appeared the universe was already huge, approximately 100 million light years across. The distances were too large to allow light to correct the differences in temperature. And at this time the most remote regions of the universe had already agreed to reach exactly the same temperature, at distances of millions of light years.

Only cosmic inflation enables us to understand how this could have happened. All the other alternative mechanisms that have been considered have turned out to be implausible in the end.

Before inflation, in the tiny bubble that was battling with the constraints of quantum mechanics, all parts were in contact with each other, like the point referred to in Calvino's *Cosmicomics*. Being able to exchange information, they all possessed the same properties, and temperature in particular was the same. Inflationary expansion propagates this homogeneity at a cosmic level and makes it become a general property of the universe. In doing so, it also magnifies out of all proportion even the infinitesimal quantum fluctuations that are present inside the primeval bubble. Billowing space outwards, it also amplifies the small disturbances that will continue to grow until they reach the dimensions of clusters of galaxies. Diffused at a cosmic level, those tiny crenellations of energy will become a thin net enveloping everything, its knots acting like seeds to produce new aggregations of matter. Those variations of density will help to thicken filaments of dark matter and will attract gas and dust around which the first stars will materialise, and the first galaxies will be formed.

From these incandescent relations, strictly determined and at the same time chaotic, between sidereal distances of the cosmos and the infinitesimal world of quantum mechanics, the material structures came into being from which dynamics and beauty were born. A world without fluctuations would not have produced stars, galaxies, planets: a perfect universe would have no spring breezes or laughing girls. We are all descended from this anomaly we call inflation, which took quantum foam to cosmic dimensions.

When the most sophisticated instruments carried by satellites demonstrate that the distribution of isotropies is exactly that predicted by inflationary models, even the most convinced critics of the new theory have to acknowledge its predictive power.

There remained, however, a huge discrepancy with the potential for provoking a new crisis of doubt, causing everything to collapse like a house of cards. Inflation necessarily entailed a universe with no local curvature, that is, one that was basically flat. The curvature of space-time depends on density, namely on its matter and energy content. For density exactly equal to the critical density the universe is flat, its local curvature is zero like that of a flat surface, and this means the expansion continues indefinitely. For higher densities the universe is closed, its local curvature is positive like that of a sphere, the expansion diminishes and the Big Bang is reversed, becoming the Big Crunch. For lower densities the local curvature is negative, like a horse saddle, and in this case as well the expansion continues indefinitely.

If inflation had really taken place, the universe could only be flat; the initial dimensions of the microscopic bubble would have been *ironed out and flattened* by the furious expansion of the first instants, and only an early universe with precisely zero curvature could have remained flat for billions of years. Any initial deviation from this condition would have been amplified to an immeasurable degree by subsequent expansion.

In other words, one of the most significant verifications of the theory of inflation could be had by measuring the local curvature of the universe or its matter and energy density. It was at this point that further problems arose.

You can get the local curvature of space-time again from fossil background radiation. We simply measure the angular diameter of the small inhomogeneities of temperature difference that amounted to no more than one hundred thousandth of a degree between one region of the heavens and another, descendants of primeval statistical fluctuations. And here the experimental data flawlessly reproduced the predictions of inflation, telling us that the universe was indeed flat. But this result was completely at odds with measurements of the energy density of the universe that seemed to indicate until the early 1990s that the universe was open, with a saddle-shaped curvature.

For several years this discrepancy proved to be the weak point of inflation theory, and provoked the objections of many detractors. Inflation needed to be discarded because it necessarily implied that the density of the universe was equal to the critical one, whereas the more accurate observations up until the mid-1990s indicated that it barely reached a third of that.

It was with the discovery of dark energy in 1998 that this argument was undermined. By observing that the recession speed of the most distant galaxies increases with time, the idea had to be accepted of an unprecedented form of energy which permeated all of space and contributed two thirds of the mass of the entire universe. At this stage the density value reached the critical point and the reason behind the universe being flat was understood. All of this became further confirmation of the inflation hypothesis.

In Search of the Smoking Gun

Despite the success of the theory and the numerous confirmations coming from experiments, there is still a small but determined group of critics who strongly oppose it.

There is nothing abnormal about this – on the contrary, it is wholly typical of the scientific method: to criticise relentlessly, to always doubt, to seek and test the weak points, and to evaluate alternatives is part and parcel of what it means to be a professional scientist.

It must be admitted, however, that in this case there was still an issue with the theory that was easy for the sceptics to focus on. Ultimately, inflation is born from a scalar field that emerges from the void with its unstable potential that propels the expansion – but up until now no one has found an unmistakable trace of the *inflaton*, the particle that is associated with this field. On the day that such a discovery is made, all doubts will be eliminated: we will have found the 'smoking gun' in the strange case of cosmic inflation. But this has not yet happened, and the search continues.

Alan Guth's initial idea was based on the conjecture that it might have been the Higgs boson that triggered everything. The phantom particle, at the time, was only a hypothesis – the key part of a theory that like many others before it could well turn out to be the result of an arbitrary conjecture. What's more, it did not predict precise values for the mass of the boson, or other properties related to it. With the Higgs boson playing the role of the inflaton, it was easy enough to explain how inflation had begun, and anything but easy when it came to discovering a mechanism that could stop it.

In reality, scientists including Guth himself quickly developed models in which various scalar fields could trigger the same mechanism. The role of blocked potential, conjectured for the Higgs boson as a false vacuum state, could be played by a potential that was weakly variable, slowly falling over time, while the primeval bubble expanded. In this vein entire families of inflationary models of different kinds were developed, the characteristics of which depended essentially on the way in which they used the hypothesis of the inflaton.

There were even some who theorised models that entailed an *eternal inflation*. Starting from the idea that quantum fluctuations of the scalar field could trigger an inflationary paroxysm beginning from a tiny portion of itself, from which point a universe was born and continued its evolution, it was possible that from the remaining matter surviving at its margins others could develop, in a mechanism of eternal inflation capable of producing that myriad of universes posited in modern theories of the *multiverse*.

Only with the discovery of the inflaton will we have, on the one hand, irrefutable confirmation that the theory is correct and, on the other, a way of discriminating between the various models that have been proposed.

When, after a search that had lasted almost fifty years, the Higgs boson was discovered at CERN in 2012, and its properties, including its mass, were measured, the debate about its possible role in the inflationary phase was immediately resumed.

This newcomer was the fundamental scalar particle, and some cosmologists still believe that it is nothing less than the inflaton itself. Others consider it to be too heavy and dispute the suggestion. Therefore a similar but lighter particle is being sought, one that could appear in some rare instances of decay produced by collisions in the LHC, or some other closely allied scalar with which it could have shared the demanding work of giving rise to an entire universe.

Opinions on this issue are quite contradictory, and the solution will only emerge from a new programme of experimental studies.

In the next few years we expect much more precise measurements to be made of cosmic background radiation, capable of clearly establishing the evanescent traces left behind by inflation. With the recent discovery of gravitational waves, it is hoped to even increase the sensitivity of the new instruments to a level that will allow us to identify fossil gravitational waves, those imperceptible fluctuations of space-time that give 'live' coverage of what was happening during inflationary growth.

Always supposing, that is, that we are not startled during the experiments we are conducting at the LHC by the discovery of a new scalar possessing all the right traits to match the identikit picture of our number-one suspect.

The Golden Age of Grand Unification

Inflation is by no means the first act to take place on our stage, though it is undoubtedly one of the most spectacular. We are not yet equipped to describe what happened in the extremely brief instants before it began, but we do know that important events took place at this time. An insurmountable wall impedes our understanding.

We can only venture various conjectures, like Plato's prisoners. In chains since childhood, their legs and necks bound, deprived of all experience of the outside world, they are unable to perceive directly what occurs outside their cave, beyond its walls. That is why they are obliged to fashion their own vision of the world from the shadows cast by it upon those walls. We scientists do something similar in order to gain an intuition of what may have happened before inflation. All we can see are shadows, and what we can imagine from them.

We can make accurate measurements at the scale of

energy that we can explore directly, by using the particle accelerators or studying the most energetic phenomena produced by the cosmos. Subsequently we extrapolate from these results to the scale of energy that we are incapable of studying directly, and we develop conjectures consistent with all the observations that we have gathered.

We are talking about the initial phase of the universe, the flicker-like duration of which was so brief as to be equivalent to Planck time, 10^{-43} seconds, corresponding to a dimension of the universe equal to 10^{-33} centimetres. At this level, space is neither smooth nor inert, but seething with virtual particles that appear and disappear at an infernal rate. A relentless quantum effervescence issues from it, a tumultuous and chaotic space filled with roughness and inhomogeneities. In these dimensions, quantum foam seethes spasmodically and fluctuates without pause. Curvature and topology in this region can only be described in probabilistic terms.

None of the current physics theories can correctly describe what happened during the Planck era, and from various hypotheses different predictions have sprung. Beyond the wall that blocks our view lie the secrets of quantum gravity, the chimera that for decades generations of physicists have been pursuing. Perhaps this insignificant region teems with unimaginably tiny *strings* that oscillate and evolve in ten to twenty-six dimensions; or perhaps space has a discrete structure organised in infinitesimal *loops*, or perhaps the stratagems conceived by nature to quantify gravity exceed what the human imagination has been able to bring to bear upon the issue thus far.

No one has yet managed to get a glimpse of periods so

close to the initial state, or to explore such ultra-small distances. We can only devise reasonable hypotheses about the phenomena that were dominant in that interval of time: we think of it as the era of Grand Unification. Fundamental forces are unified into a single field: a unique primeval superforce governs the insignificant piece of foam that will become our universe.

The whole world in which we live is held together by forces that we can classify in decreasing order of intensity. The first on the list is strong nuclear interaction, the force that holds quarks together to form protons and neutrons, and assembles them into the nuclei of various elements. From here comes the energy that is released by nuclear weapons, or that which keeps the stars sparkling. The weak force is more timid, and significantly less conspicuous. It only acts at subnuclear distances and rarely occupies centre stage. It appears in some radioactive decays, seemingly insignificant but in reality crucial to the dynamics of the universe. The electromagnetic force keeps atoms and molecules together and regulates with its laws the propagation of light. Gravity is far weaker than these other forces, even though it is the most popular. It comes into play every time a mass or energy is present, and pervades the entire cosmos, regulating the movements of the smallest asteroids in the solar system as well as the most gigantic clusters of galaxies imaginable.

Today, in the ancient, cold universe that we inhabit, these forces act separately and have different intensities and radii of action. But what we have been able to verify, through innumerable experiments, is that all of this varies with density of energy. With its increase it seems that a principle of justice and equality is triggered, whereby 'The strong will be less strong, and the weak less weak'. The strong force diminishes in intensity, and the same happens with the electromagnetic force. On the other hand, the intensity of the weak interaction increases to such an extent that it is possible to predict where the three curves will converge: the energy that is required to fuse them into a single force.

In all of this, gravity remains somewhat apart: it is so weak that we are unable to measure variations in its intensity at the levels of energy explored thus far, but it is only natural to take it into consideration at this stage.

This early period in the evolution of the universe is called the Planck era, and is dominated by a superforce that unifies the four fundamental forces. It is comparable to looking back to a kind of golden age of holy alliance between men and gods, dwelling together and sharing loves and jealousies.

In the extremely small and heat-dominated early universe there are elegant and perfect symmetries that break down one after another as everything cools.

A first dramatic separation occurs precisely in the Planck era, when gravity dissociates itself from the other forces. Immediately after that, another step change sees the strong force separate from the electroweak force.

Our history is already launched before inflation produces the Big Bang: in a tiny area of the void, the field of a superforce gradually goes though transformational phases, breaking the symmetries that separate various interactions from each other. The subsequent crystallisation of the primordial field will fill our world with the four fundamental interactions and change everything at a stroke.

Unlike for the first two, for this subsequent rupture of symmetry that completely separates the weak from the electromagnetic force we have gathered unequivocal data that allows us to tell a detailed story. We have been able to study it in the laboratory, reproducing it at CERN with the discovery of the Higgs boson, the principal protagonist of what happened 10^{-11} seconds after the Big Bang, and the subject of our next chapter.

Day Two: The Delicate Touch of a Boson Changes Everything, Forever

As soon as it exits the inflationary phase, the incandescent universe already has all the matter and energy it needs, but if we were able to look at its interior we would be unable to recognise anything familiar. We would see a kind of formless gas made up of minute particles, indistinguishable from each other, all devoid of mass and flying at the speed of light. The whole thing would present itself like a perfect object, homogeneous and isotropic, the same at every point and from every angle. There is no aggregation, and not the slightest non-uniformity.

If it were not expanding at an enormous rate, it could be confused with the ideal representation of Parmenidean being: in every respect identical to itself, symmetrical through every rotation, totally devoid of any flaw or imperfection. This is the reign of uniformity and perfection, governed by symmetry and characterised by a combination of simplicity and elegance. If nothing startling had arrived on the scene to disrupt this seemingly immutable harmony, then nothing could have been born from this perfect object. It would have been a sterile universe, an enormous waste of effort, forever lacking the light of the Moon and the fragrance of flowers – a sad, featureless, desolate place.

We are close to the moment in which the last and perhaps the most significant transformation decreed by destiny will take place.

Having got over the euphoria of inflation, the expansion continues, impelled by an energy that is seething within its interior. As it gets bigger, the universe cools down, and in doing so triggers reactions which radically alter its dynamics.

We have reached one hundredth of a billionth of a second after the Big Bang, and from this moment onwards things become clearer. Ever since we discovered the Higgs boson and were able to measure its mass, this part of the story has kept few secrets.

The newborn universe is already pretty imposing. It has reached the not inconsiderable dimension of a billion kilometres. Suddenly, when the temperature decreases below a certain threshold, the Higgs bosons which until a moment ago had been roaming freely, now begin to congeal and crystallise. At these temperatures, which are freezing for them, they are unable to survive and are hidden away in the comfortable sepulchre of the void. It will require a great deal of patience before they are seen again. It will take 13.8 billion years in fact, before on planet Earth someone will manage to engineer collisions of energy of such magnitude as to bring them back to life, if only for a fraction of a second, an amount of time sufficient for them to leave indelible traces of their presence.

The field associated with them acquires a specific value that radically transforms the properties of the void. Many elementary particles are subject to a sizeable interaction while crossing it, and as a result their speed diminishes, allowing them to acquire a mass; others that travel without being disrupted remain impervious and can continue to move at the speed of light.

With the Higgs field the perfect symmetry that had characterised the primeval universe fractures and the weak interaction completely separates from the electromagnetic force. Some particles become so heavy they are rendered unstable, and immediately disappear from the rapidly cooling universe. Others will crucially acquire mass but retain a lightness that will prove fundamental for soon moving to a quite special organisation of matter.

The newcomer, the Higgs field, acting with a light touch, has constructed multiplicity (variety and complexity) following a clear and simple rule. The elementary particles which remain as if entangled in the field, distinguish themselves from one another depending on the intensity of the interaction, and by so doing they end up acquiring irreversibly different masses. Their gentle operation resembles that of the demiurge in Plato's *Timaeus*, the master craftsman who, through the mediation of numbers, converts the formless pre-existing matter into something dynamic and vital.

Everything will be born from this delicate touch that crucially changed things forever. But we mustn't get ahead of ourselves: it is still early. The second day has just ended and only 10^{-11} seconds have passed.

The Enchantment of Narcissus

When you see the canvas for the first time, you can't help but be spellbound before the perfect circle that holds the two figures: that of the elegantly dressed youth leaning over the water, and his reflected image that he is ecstatically admiring. Caravaggio's solution to presenting the myth of Narcissus is one of pure genius. It is based on the classic version of the story in Ovid's *Metamorphoses*. Having rejected the advances of the nymph Echo, an exceptionally beautiful young man is condemned to fall madly in love with the image of the one person he can never possess: himself. In the painting we see him reaching out his left hand towards his reflected image, hoping to touch his beloved but able to do nothing more than wet his fingers as they break the water's surface. The circle that frames the two figures emphasises the reflection symmetry which unites them.

This celebrated painting in the Palazzo Barberini in Rome is one of many masterpieces that have used symmetry as a key to explaining the nature of beauty.

The literal meaning of the Greek word from which *symmetry* derives is 'with proper measure', recalling the concepts of proportion and harmony that were accorded so much space in the aesthetic and philosophical discourse of antiquity. For the Greeks and Romans, for a work to be beautiful it must necessarily contain symmetry, with elements and volumes in mathematical relationship with each other.

Central symmetry, of the kind determining the regular distribution of the segments of an orange, or the arms of a starfish, was used widely in the classical world; one need only think of the cupola of the Pantheon, or of the Temple of Hercules Victor in Piazza Bocca della Verità in Rome.

Despite retaining connection with this original tradition pertaining to recurrent repetitions of forms and figures, to transformations by translation and rotation, the modern understanding of symmetry has added more recent elements. From this new understanding of the principle, some of the jewels of the Renaissance were produced, such as Michelangelo's cupola for St Peter's Basilica, or Donato Bramante's wondrous *martyrium* in the church of San Pietro in Montorio.

The modern notion of symmetry has made possible a mathematical formalisation that has found many applications in the sciences. For physics in particular, symmetry is not just a concept that implies constancy and elegance of relations. It is a real tool of investigation that has enabled the construction of new laws of nature. And this is all thanks to Emmy Noether, perhaps the greatest mathematician who has ever lived.

The young German researcher had to labour for years before she was permitted to teach at a university: she was a badly paid and barely tolerated teacher when in 1918 she managed to formulate the relation that would change the face of contemporary physics forever. Noether's theorem establishes that for every instance of continuous symmetry in the laws of physics there is a corresponding *law of conservation*, that is to say a physical quantity that is measurable and which remains unchanging.

The most common examples of this are the symmetries that give rise to the principles of conservation in classical mechanics. If a system follows laws of motion that do not change when the frame of reference shifts – symmetry of spatial translation – then the quantity of motion is conserved. If the same invariant result is obtained for translation of the time axis, then energy is conserved. If the same happens when rotation is involved, then angular momentum is conserved. And so on.

In contemporary physics this relation between symmetries,

transformations and the conservation of physical quantities will become generalised. The invariability of some physical properties in a system subjected to transformations will facilitate the discovery and formalisation of the relations that will provide the basis for a new conception of matter. From this, the principles of conservation of physical quantities with peculiar names will arise and prove crucial to the description of the most minute components of matter: *strangeness*, *isospin*, *lepton number*, and so on.

The concept of symmetry will assume a more general application, according to which it will become common to speak of symmetries that are *continuous or discrete*, *local or global*, *exact or approximate*: each term constituting a fundamental tool for understanding the dynamics of elementary particles and their fields. Without Emmy Noether's remarkable contribution, none of this would have been possible.

The culmination of this effort will be the development of the Standard Model of elementary particles, a truly monumental achievement which encompasses the most accurate and precise description of matter that we currently have.

It is the most successful theory in contemporary physics, and explains matter through a limited number of components: six quarks and six leptons, each one organised into three different families. The twelve particles of matter combine together, or interact with each other, exchanging other particles which transmit the forces: the *photon* that carries electromagnetic interaction; the *gluons* that transmit *strong* interaction; the *vector bosons* W and Z that propagate weak interaction. The particles of matter, leptons and quarks, have half-integer spin (1/2) and constitute the family of *fermions*, while the particles that carry interactions have integer spin (0, 1) and form the family of *bosons*. With this limited list of ingredients we can construct all the known types of matter, both the stable ones that populate our daily lives and those exotic and ephemeral ones produced in our accelerators, or in the extremely high-energy processes at the heart of stars or manifest during cosmic catastrophes.

The theory immediately enjoyed remarkable success, thanks to its tremendous predictive power. From the outset of its formulation in the 1960s it predicted new particles that were then regularly discovered, and it made it feasible to calculate with great precision new parameters that when measured were found to match the predictions – sometimes accurate to within ten decimal places.

The architrave of the Standard Model is provided by the unification of the electromagnetic with the weak interaction, which thus become two different manifestations of a single force: the *electroweak* interaction.

Once again, everything emerges from a symmetry. The first to catch a glimpse of it was Enrico Fermi, when, barely in his thirties, he intuited that behind an apparently marginal phenomenon – radioisotopes that decayed emitting electrons – a new kind of fundamental force was hidden. Fermi conjectured that a strong formal analogy existed between the new interaction and electromagnetism, and deployed it to construct a description of the new force and to calculate the coupling constant.

For many years the force will be known as 'Fermi's interaction'. It will eventually become what we now know as 'weak' interaction, the change of name drawing attention to the small value of the constant that determines the intensity of the force and is still called 'Fermi's constant' in honour of its discoverer.

The young scientist's innovative idea opened the way to that unification of the electromagnetic and weak forces which, thirty years later, would constitute the basis of the Standard Model of fundamental interactions.

In 1865, James Clerk Maxwell published the equations that provide the basis for the theory of the unification of electrical and magnetic phenomena: electromagnetism was born. A century later the story is repeated. In the mid-1960s Steven Weinberg, Sheldon Glashow and Abdus Salam are able to formalise the new theory with a key contribution from Gerard 't Hooft. Electromagnetism and the weak force are two different manifestations of the same interaction that from now on will be known as the electroweak force.

The discovery by Carlo Rubbia in 1983 of W and Z, the vector bosons predicted by the new theory, will mark the definitive triumph of the Standard Model.

Beneath the superficial appearance of success, however, there lay hidden a deep crack, a weakness that was intrinsic to the theory and could have brought down the whole edifice by undermining its main supports.

It all came from the simplest of questions, namely: how is it possible that the two interactions, so different from each other, are nevertheless manifestations of the same force? The electromagnetic force has an infinite radius of action, whereas the weak interaction only manifests itself across infinitesimal subnuclear distances.

A general law of physics states that the radius of action of a force is inversely proportional to the mass of the particle that carries it. The photon has a zero mass, and therefore the electromagnetic interaction may reach the most disproportionate distances. At the other end of the scale, W and Z are very massive, weighing in at the equivalent of 80–90 protons, and their radius of action is tiny. The weak force acts within the interior of the nuclei, which is why we had failed to notice its presence until recently.

But how can the photon, which is devoid of mass, mediate the same electroweak interaction carried by W and Z? What is it that really differentiates W and Z from the photon? What exactly is the *thing* that we call mass?

The Beauty of Broken Symmetry

Castelfranco Veneto is one of the hidden treasures of Italy. It has preserved the original structure of the walled city, originating from the castle that defended it. Its cathedral, erected as it should be in the centre of the city, is an attractive neoclassical building. It is a church of modest proportions – nothing like those imposing ones found elsewhere. But no sooner has one entered it and approached the Costanzo chapel, to the right of the presbytery, than one is amazed. Its altar enthrones Giorgione's *Pala*, the masterpiece of the painter whose home can still be visited in a small square nearby.

Despite the brevity of his life, Giorgio Barbarelli, for that was his real name, managed to bequeath to the world unforgettable works. Barbarelli was only twenty-five years old when he began painting the *Pala*, commissioned by Tuzio Costanzo – a *condottiere* from Messina, hired by the *Serenissima*, the army of the Venetian republic, to command its forces. He wanted an altarpiece for the mortuary chapel dedicated to his son, Matteo, who had died of malaria during a military campaign near Ravenna, aged just twenty-five.

In carrying out the commission, Giorgione decides to break with tradition. The masters before him – from Piero della Francesca to his own teacher, Giovanni Bellini – had always placed their figures at the centre of an ideal composition, within a perspectival and sophisticated game that sometimes echoed the lines of the church in which it was to be set. Giorgione keeps the strong iconographical structure of the pyramid, at the apex of which he positions the Madonna and child, but decides to split the perspective and open it towards the outside. The extraordinarily high, unnatural, almost metaphysical throne contrasts with an affectingly serene landscape, of countryside and hills bathed in suffused light. Both in the rendering of the figures and in the treatment of the background there is a celebration of the triumph of tonal Venetian painting, the distinctive touch that distinguished the Venetian from the Florentine school, showing that 'painting without drawing' referred to by Vasari in his *Lives*. It is a masterly technique employing layers of overlapping colour, avoiding any abrupt transition between light and shadow, enveloping all the edges in a suffused and delicate chiaroscuro.

This great picture has a double axis of symmetry: top–bottom and right–left. A capacious dark-red velvet fabric provides the backdrop for the earthly realm, with its regular, orderly, chequerboard floor which supports the base of the throne and the two lateral figures. Above it the celestial world is offset by a landscape of haunting melancholy, with the Virgin Mother at its centre.

The perfect symmetry is broken in the top section by

the figure of the child, sitting on the Madonna's right knee, absorbed in awareness of his destiny. Below, the two figures adopt the same posture and are placed in a perfectly symmetrical position in relation to the central axis of the painting; both figures look straight into the eyes of the viewer and bring him into the intimate interior of the painting, but it would be hard to overstate the contrast between them. On the right there is Saint Francis, looking listless, unarmed, in the modest clothing in which he has come humbly dressed to Damietta carrying his message of peace to the Sultan of Egypt, al-Malik al-Kamil; on the left the ostentatious, gleamingly armour-suited Saint Nicasius, a warrior-monk of the Knights Hospitaller of Saint John of Jerusalem. He had been fighting as a Crusader in the Holy Land, and having been captured at the Battle of Hattin, was decapitated in the presence of Saladin – the uncle of the sultan who, years later, would enter into peaceful dialogue with the saint of Assisi. Nicasius is carrying the standard of the Jerusalemites, the flag with the cross that will become the insignia of the Knights of Malta, and the pike which supports it is the last and most important element which disrupts all possible symmetry: invading the celestial space, it ruptures the division between the two worlds, and ultimately breaks down with an aggressive diagonal the vertical order of the composition. Here, in a single painting, is a deliberate breaking of symmetry devised with absolute mastery. It is a masterpiece of both beauty and innovation.

The appeal of 'broken symmetry' can be found in many works of art. The orderly rhythm of perfect symmetry tends to pacify and reassure, but it risks ultimate blandness: it does not elicit emotion, because it fails to surprise. The effect of the break is unsettling but also intriguing; it pushes us beyond the limits of our certainties, to investigate where we are being taken by this skewing of equilibrium. For an instant we seem to hesitate, we are overcome by trepidation generated by the unexpected innovation and the risks that accompany it; then the artist reassures us and returns us to familiar territory. Just as when, having pursued a variation upon a dominant theme in a symphony and having sensed that we were losing our way, equilibrium is restored by its rediscovery during the finale, in a kind of concluding embrace. These are techniques used with great mastery by eminent painters, and by composers of genius such as Bach and Mozart. From this irregularity the secret of the insuperable fascination exerted over us by masterpieces originates – from the Leaning Tower of Pisa's departure from the vertical to the intriguingly asymmetrical smile of the Mona Lisa, and all the way down to the gilded bronze sculptures of Arnaldo Pomodoro: those polished and flawless spheres, the creatures of magical mathematical relations, which he lacerates and breaks apart, exposing their tormented inner workings.

If, in the field of art, dismantling symmetry is a deliberate act that provokes fascination and astonishment, how should we explain the fact that nature seems distinctly inclined to resist succumbing to the same process?

The universe that emerges after the phase of inflation is in a state of perfection. The laws of physics that regulate it are marvellously symmetrical. Why does such a perfect mechanism shatter?

To understand the role for physics of spontaneous symmetry breaking, we can resort to a mechanical example: a pencil resting on its point on a flat surface. The initial state of the system is perfectly symmetrical. The pencil can rotate on its axis and the laws of physics are the same, because the gravitational field is symmetrical for those rotations around the vertical axis. In other words when the pencil falls flat to the surface, it will do so in any direction. The symmetrical state is unstable, and left alone the pencil will fall. On the horizontal plane the pencil is stable, but it has broken the rotational symmetry of the gravitational field because it has opted for a particular direction. Falling on the flat surface the pencil may have lost energy and symmetry, but it has gained stability and multiplicity.

Something of this kind occurred in the early universe. The initial incandescent state had a high level of symmetry, but it was unstable; as it cools down it loses symmetry but acquires stability.

But what was the state of lower energy into which this universe was placed? What mechanism could cause the electroweak symmetry to spontaneously rupture?

The problem was there from the first growing pains of the electroweak theory and various kinds of solution had been proposed, none of them ultimately convincing. The solution arrived in 1964, thanks to three thirty-something scientists: the Belgians Robert Brout and François Englert, and the Englishman Peter Higgs.

Once again, a few young men propose a novel idea that's outside established frameworks, and no one takes it seriously at first, precisely because it is so revolutionary.

If the equations of the two interactions are the same, the symmetry can only be broken by the same medium in which it was propagated. That is, the void. In other words, it is the void that is responsible for 'breaking the symmetry', because the void is not ... void. Since time immemorial every corner of the universe has been occupied by a field. It is the Higgs field, produced by the Higgs boson, an elementary scalar particle that must be added to the particles of the Standard Model. Only by following this reasoning can we explain how the electromagnetic force and its weak counterpart behave in such different ways as to dispel any notion that they may be even distantly related.

Instead, in the small incandescent early universe, the Higgs field was experiencing an excited state that made everything perfectly symmetrical. As the temperature drops, it is frozen into a state of equilibrium at lower energy that breaks the initial symmetry. W and Z acquire mass because they remain heavily entangled in the field that holds them captive, while the photon continues to wander everywhere, without mass: it is oblivious to this new development because the field does not even tickle it.

A similar mechanism explains why leptons and quarks have such different masses. They are also born democratically without mass. It is the Higgs field that selects and distinguishes the heavy from the light particles, determining which will be which. The greater its interaction with the field, the larger the mass of the particle.

Everything seemed to be elegantly solved ... except for one small detail. Did the Higgs field really exist? Who could prove this elegant solution was the one actually chosen by nature? If the field existed somewhere, the particle associated with it needed to manifest itself. So started the great race to discover the Higgs boson.

The Discovery of the Higgs Boson

It took almost fifty years before it became possible to verify that the Higgs mechanism was in fact responsible for breaking electroweak symmetry. That is how long it took to find the most elusive particle in the history of physics.

The theory could not predict what kind of mass the Higgs boson possessed, which effectively meant that it could be hiding anywhere. For decades, scientists from all over the world made superhuman efforts to capture this new particle, with no satisfactory results. We now realise, after its discovery, that these failures happened because the Higgs was so heavy that the energy generated by the accelerators used prior to 2010 was not sufficient to produce it. The turning point came with the construction of the Large Hadron Collider, the huge scientific infrastructure located at CERN in Geneva.

Particle accelerators have taken on the role of time machines: they can plunge us back billions of years, allowing us to study the phenomena operating at the birth of the universe. In the collisions that are produced, one strikes the void and it is possible to extract from it particles of matter. It constitutes an application of Einstein's celebrated relation of equivalence between mass and energy, $E = mc^2$. When beams of particles collide with each other, the energy generated by the collision may be transformed into mass: the greater the energy, the heavier the particles that we can produce and study in detail. They are consequently *manufacturers of extinct particles* which bring back to life, for fractions of a second, forms of matter that had vanished immediately after the Big Bang.

The LHC is currently the world's most powerful particle

accelerator in operation. Two beams of protons made up of thousands of individual small packets circulate in opposite directions in an empty tube with a circumference of 27 kilometres. In every packet there are concentrated more than 100 billion protons which are accelerated by very high electrical fields, while powerful magnets bend their trajectories to keep them in orbit and bring them into collision. The energy generated by the LHC is 13 TeV (or tera electron volts, which is 13,000,000,000,000 electron volts) but as the protons are made up of quarks and gluons, their collisions are rather complicated and only part of the available energy, a few TeV, may be converted into massive particles. The heavy protons, moreover, lose only a small amount of energy due to radiation and hence it is easy to propel them towards higher energies. This is why proton colliders are the machines most adapted to the discovery of new particles.

The electron-positron colliders have a complementary function. Given that the particles at stake are pointlike, their collisions are very simple and the whole energy of the collision may be exploited to produce particles. These are the ideal machines for attaining measurements of unparalleled precision, and for attempting to discover new particles indirectly, through the study of subtle anomalies.

The disadvantage of electron detectors, however, is that they don't allow the attainment of extremely high energies. Light particles such as electrons move through circular orbits radiating large numbers of photons, and as a result they allow a significant fraction of their energy to escape. This loss grows drastically with the increase in energy and ultimately constitutes an insurmountable barrier that limits the direct discovery of new particles.

The energies that are released by the collisions of particles produced by the accelerators are really insignificant when compared to the scale of our everyday existence. But there, concentrated in the infinitesimal space in which these impacts take place, the extreme conditions that have not occurred since the Big Bang may be recreated. In these collisions, hidden among a myriad of well-known and rather more conventional phenomena, the special events were produced that made it possible to identify the Higgs boson.

This outcome was made possible thanks to the work of two distinct groups of researchers known as ATLAS and CMS, each consisting of thousands of scientists. The decision to have two experiments is almost obligatory when it comes to discovering new particles. The signals that are being looked for are so rare, and the chances of error so great, that only by having two independent experiments based on different technologies and constructed by independent groups of scientists is it possible to attain the kind of certainty that will not be treated as a false alarm.

ATLAS and CMS were conceived so that they would work in ways that were completely independent of each other, and this has generated a degree of fierce competition between them: if one of these groups manages to be the first to discover a new state of matter and the other gets there late and is reduced to merely confirming the findings of their competitor, the whole glory of the discovery will go to the first group. For this reason no one working in either of these groups sleeps easy: the nightmare that something might go wrong, or that the other group will get to the finishing line first, is always lurking just around the corner.

Owing to an almost incredible set of circumstances, what

happened instead was that both the experiments worked perfectly and the teams arrived at the finishing line together. They simultaneously identified the first signs of the presence of the Higgs through their data, and when the signal became so strong as to dispel all doubts and caution, they were able, in 2012, to announce the discovery of the particle together. The newcomer has a mass of 125 GeV (giga electron volts, or a billion electronvolts) and has turned out, in all respects, to be remarkably similar to the Higgs boson predicted by the 'Boys of '64'.

With this result the Standard Model celebrated another triumph – one that was recognised with the 2013 Nobel Prize awarded to François Englert and Peter Higgs, the two surviving members of the trio of young scientists who together were the first to hypothesise its existence.

Who Broke the Symmetry between Matter and Antimatter?

Now that we have discovered the new particle, things have become clearer: we can understand better when the transition took place, and outline the contours of the mechanism of the spontaneous breaking of electroweak symmetry.

A time X depends on the mass of the Higgs boson corresponding to a specific temperature reached by the primeval universe 10^{-11} seconds after the Big Bang. Starting from that moment, the electromagnetic interaction completely separates from the weak interaction, thus initiating a long process that will lead all the way to us. Just as the pencil falls to the surface of the table, the universe has lost its symmetry but has gained stability and multiplicity. Everything that surrounds us, the wonder of its infinite variety of forms that is still a source of astonishment to us, could not have emerged if the fearful symmetry in which it was held had not been ruptured. The Higgs boson was the kiss that broke the spell imprisoning the princess in the deathly perfection of absolute uniformity. From that irregularity, from that small fundamental *flaw*, everything was unleashed.

Today it is possible to describe the potential associated with the new scalar field and better understand the mechanism that had such a fundamental role in building the material structure of the universe.

Perhaps it is in that magical moment that we will also find the key to resolving the mystery of antimatter, because with the discovery of the Higgs boson, new hypotheses are emerging.

The first idea relating to antimatter dates back to 1928 and comes about almost by chance – from the calculations of Paul Adrien Maurice Dirac. The young English scientist, just twenty-six years old at the time, was attempting to formulate a theory that would explain the behaviour of subatomic particles at high energies. In order to do this, he needed to reconcile the description of the particles furnished by quantum mechanics with the transformations due to relativistic effects. When he set up the relativistic equation for the motion of electrons, he realised with astonishment that the same equation was also valid for *positive electrons*. What initially seemed like a pure formal coincidence was soon considered to be the discovery of another of nature's fundamental symmetries. Relativistic quantum mechanics tells us that, for every particle endowed with a charge, there must be another particle possessing identical mass but with an opposite charge: what we call an *antiparticle*. The idea that the elementary components of an *antiworld* actually existed was so bizarre that, from the outset, no one took it seriously. Things changed when another young scientist from Caltech, the twenty-two-year-old Carl David Anderson, focused his attention on some strange traces that appeared in the detector he was using to study cosmic rays. After an endless number of controlled experiments, his conclusion was unequivocal: he was dealing with particles that had the same mass as the electron but with a positive charge. It was thus that the first *positrons* were discovered. While undeniably rare, antimatter was a real component of our material world.

Ever since, with impeccable regularity, as the catalogue of new particles gradually became enriched, so in parallel did that of their partners of opposite charge.

Antimatter has now become quite common. It is produced to use, or in order to study its properties, in many particle accelerators, but it is also used in routine clinical procedures in many hospitals. The most common example is PET, or positron emission tomography, a diagnostic technique that allows the reconstruction of functional images of internal organs, starting from the annihilation of positrons and electrons.

One of the properties that has most touched the collective imagination is precisely this characteristic: particles and antiparticles that come into contact with each other are transformed into pairs of photons with a total energy equivalent to the mass of the initial system. This highly efficient transformation of matter and antimatter into energy has inspired whole strands of science fiction.

In fact, no reaction can compete with the process of annihilation. The energy that could be produced by combining a kilo of matter with a kilo of antimatter is seventy times greater than that generated by causing the nuclear fusion of a kilo of liquid hydrogen, and 4 billion times higher than that produced by the combustion of a kilo of petrol. The only problem is that no one has yet come up with an efficient mechanism for producing large quantities of antimatter. The particle accelerators produce infinitesimal quantities, and only then with a truly enormous expenditure of energy and materials. It has been estimated that in order to produce 10 milligrams of positrons we would have to spend 250 million dollars. In short, antimatter would cost 25 billion dollars per gram, thus making it by a very long way the rarest and most costly material on Earth. Perhaps needless to say, the prospect of constructing spacecraft with engines powered by antimatter – like the *Enterprise* in *Star Trek* – is hardly feasible.

Since its earliest formulations, the concept of antimatter has always been accompanied by an issue for which physics has not been able to find an answer: if the equations are symmetrical and describe the behaviour of matter and antimatter in an equivalent way, why is our world so dominated by matter? It seems natural to assume that, at the end of the inflationary phase, the excess of energy would have extracted equal amounts of matter and antimatter from the void. But antimatter seems to have completely vanished from the universe around us. So where did it go?

Thousands of scientists are busy looking for an answer right now, taking different roads in order to do so. The first assumes that great quantities of antimatter may have escaped into regions of space that we have not yet explored: whole worlds made up of antimatter, huge galaxies of antiprotons and positrons, that have until now eluded all attempts to observe them.

The second line of research conjectures that everything may be due to a subtle difference in behaviour between matter and antimatter, a small anomaly which breaks the original symmetry and is the key to everything. Detailed studies have been conducted, and mechanisms in fact found that show a very slight prevalence for matter in the processes of decay of particles and antiparticles. These differences were foreseen by the Standard Model, but the preference attributed to matter turns out to be too small to explain the excess that we observe all around us.

Finally, in the last few years a new hypothesis has been proposed. Everything might have been determined by something very special that happened precisely when the Higgs boson occupied centre stage and broke that perfect symmetry prevailing in the original universe. A slight preference might have been enough to produce a coupling with particles rather than antiparticles, thus producing the material universe that surrounds us.

Still other theories have been mooted. That the asymmetry may have been born, for instance, from the modalities of the electroweak phase transition. According to the speed at which this dramatic change happened, a local anomaly might have become a widespread general property in the new system, and at that point the bifurcation between matter and antimatter would have been created. Our material universe would then have taken the path of matter, definitively abandoning that of antimatter.

To study these phenomena in detail it will be necessary to produce tens of millions of Higgs bosons and to closely measure the most minute properties, looking out for any possible anomaly. These are the investigations being carried out with the aid of the LHC, as the capacity of the machine for producing collisions gradually increases. But we may need to build an even more powerful accelerator in order to understand exactly what happened. Powerful enough to disturb the Higgs field and thus to reconstruct all the stages of that fateful transition, allowing us to study its distant behaviour from the comfort of that equilibrium in which it has rested now for billions of years.

The Deepest Symmetry of All

Under the term *supersymmetry* there lies in reality a complex family of theories, united by the hypothesis that every known particle has a supersymmetric partner: a particle that effectively resembles it in every way except that it is much heavier and has a spin that differs by ±1/2. Therefore supersymmetric bosons with whole spin (0, 1) correspond to ordinary fermions with a half-integer spin of 1/2, while ordinary bosons are complemented by supersymmetric fermions. In this *superworld* the fermions have the task of carrying interactions while the bosons constitute matter.

The theory predicts that this enhanced form of symmetry was also broken in the first moments after the Big Bang. In other words, supersymmetric particles used to fill the incandescent environment of the primordial universe in equal proportion to ordinary matter. But rapid cooling due to expansion produced a mass extinction. Incapable of surviving at lower temperatures, supersymmetric particles disintegrated almost immediately into ordinary matter, and this explains why we find none at large today.

However, it is quite possible that there have been exceptions. The theory contends that it is possible for stable supersymmetric particles to exist, in other words ones that cannot decay into something else. These heavy particles, which could only interact weakly, could conceivably build huge masses capable of exerting intense gravitational pull. If this proved to be the case, it would allow us to understand the origin of the dark matter that holds together galaxies and clusters of galaxies. These vast concentrations of stable supersymmetric particles could represent the fossil residue of the early universe in which supersymmetric matter dominated the world.

The fascination exerted by SUSY – the acronym used for the set of supersymmetric theories – also comes from the fact that a simpler scenario for the unification of fundamental reactions would follow from the theory, and that there would be a special place in it as well for the Higgs boson. The particle discovered in 2012 might in reality be the first of a whole family of super-Higgses, and supersymmetry would allow us to better understand why it has a mass of 125 GeV. Virtual supersymmetric particles would protect it from the instability that besets, through quantum effects, a scalar of that mass, building around it a kind of armour.

For the theory to be verified, it is not enough that it happens to be elegant and enjoys a considerable popularity among theoretical physicists. Such peculiar particles actually need to be identified in the data of some experiment or other, and so far this has not happened. It is therefore quite possible that the theory is wrong. Or perhaps the supersymmetric particles are so heavy that we cannot manage to produce them even with the current LHC. If this were the case, we could still detect their presence through their *virtual* effects. Ultramassive particles can float like ghosts around the known particles and interfere with them through mechanisms predicted by the Standard Model. From this process, anomalies might arise that could be recorded by our detectors, thereby constituting an important 'indirect' discovery of new physics.

This is how the hunt for supersymmetry is continuing unabated, on several contemporary fronts. By utilising the increase in energy of the LHC to 13 TeV since 2015, it is hoped that it will be possible to produce those massive particles that have escaped detection by every investigation to date. At the same time the cousins of the Higgs are being looked for in the region that has already been explored in the search for the scalar of the Standard Model. What has been produced so far is not enough, because what we are looking for are particles with very distinct characteristics. The supersymmetric cousins of the Higgs have ways of production and decay that are peculiar, and therefore require the elaboration of highly specific strategies. We also need a great deal of further data, since they may be particles more difficult to produce and much rarer to find.

Independently of all this, the studies on the Higgs boson at 125 GeV continue. The Standard Model predicts all its properties with great precision. Up to the present day what we have witnessed is consistent with the predictions, but our degree of precision is limited by the small number of bosons we have managed to produce and trace. For many processes of decay the level of uncertainty of our measurements is still too high and could be concealing the anomalies anticipated by SUSY.

Accurate, patient, and systematic investigative work continues at the LHC. No stone will be left unturned in the search for unequivocal signs of supersymmetry, in the secret hope that the recently discovered Higgs boson can function as a portal opening onto a new physics; and that what happened in 2012 will prove to be the first link in a long chain of discoveries.

Accelerators of the Future

Physics is undergoing a process of radical transformation. Now that we have found the last particle to have heeded our call, the Standard Model of fundamental interactions is complete. But at the very moment when a new triumph of the theory is being celebrated, everyone is nevertheless acutely aware that the list of phenomena for which it provides no explanation is so long as to be frankly embarrassing.

We do not yet understand the exact dynamics of inflation; nor has it been possible to coherently unify the fundamental forces, including gravity. We are completely in the dark as to the mechanisms that led to the disappearance of antimatter, not to mention the phenomena that might explain dark matter and dark energy.

Everyone is aware that sooner or later the Standard Model will have to be reconfigured. It will probably become a particular case of a more general theory that's capable of opening up a new and more complete description of nature. The beauty of the labour of research is that nobody knows just when this shift will take place. Any day could be the right one: perhaps all that's needed is for a new state of matter to jump out from the latest analysis of data from the LHC; or maybe years of further attempts will need to be made, possibly even with a new generation of as yet unbuilt accelerators.

So while the work proceeds, the instruments of the future are already being planned. The timescale for the development and implementation of a new accelerator can be measured in decades. The first discussions of the LHC began midway through the 1980s: the machine itself was completed in 2008. If we want a new machine to be up and running in 2035–40, we need to act right now. It is no coincidence that at the start of 2019 CERN published the report that describes the FCC project. The acronym stands for future circular colliders, the accelerators that will be the descendants of the LHC.

FCC is an international study group with the remit to produce a project, define the infrastructure, and estimate the costs for a collider 100 kilometres in length to be built at CERN. The project anticipates, as a first step, a particle accelerator that will produce collisions between electrons and positrons, FCC-ee, that will later be converted into a proton–proton machine, FCC-hh, following the successful design already used at CERN with the LEP (large electron-positron collider) and LHC series.

The proposal, begun in 2014, immediately gathered enthusiastic support from the international community. The work involved over 1,300 physicists and engineers, belonging to some 150 universities, research institutes and various industrial partners. The result of the study is a detailed report that constitutes the basis for defining the new European strategy in the field of particle accelerators.

The decision to build this unprecedented facility was unanimously endorsed by CERN's governing body on 19 June 2020. Realistically speaking, the construction of the FCC-ee might begin in 2028 and become operational before 2040. The proton machine would be much more complex, and would need years of development for the industrial production of the magnets. The start of FCC-hh could be somewhere between 2050 and 2060.

To summarise the current situation: decisions are already being taken now that will determine the boundaries of scientific research for the rest of the century.

From a research point of view, the combination of two sequentially linked accelerators is by far the optimal configuration. It appears to afford a kind of pincer movement so as not to let the new physics escape, wherever it might be hidden.

The electron–positron machine provides the ideal environment to carry out precise measurements of the Higgs and of the fundamental parameters of the Standard Model. It is anticipated that the new accelerator would function at first at 90 GeV to produce a huge number of Z bosons, and then move to 160 GeV to generate W pairs, before rising to 240 GeV to produce millions of Higgs particles in association with Z, and ultimately to 365 GeV to create pairs of top quarks.

The new particles that would help us to explain dark matter, or new interactions that would transport us to the hidden dimensions of our universe, could be discovered in an indirect way through the most incredibly accurate measurements of the parameters of the Standard Model ever imagined.

If precision is not enough, we will be able to call on brute force. With the 100 TeV of FCC-hh energy, it will become possible to explore on an energy scale seven times higher than is achievable with the LHC. Whatever new state of matter appears, with a mass ranging from a few to some tens of TeV, it will be identified directly and we could also understand whether the Higgs boson is indeed elementary or has an internal structure. It will also be possible to study details of the spontaneous breaking of electroweak symmetry – details that could be decisive for understanding the prevalence of matter in the world.

The project comes with significant costs. Nine billion euros will be needed to dig the tunnel and equip the electron machine. An additional 15 billion will be needed to manufacture the powerful magnets needed for FCC-hh. However, if we take into consideration the time span over which this investment will be made, and the global financial contributions that it should attract, the project definitely seems to be viable. What's certain is that with FCC, Europe is launching a challenge that will be centre stage in the worldwide debate about accelerators of the future.

The United States, undeniably the leader in this field until a few decades ago, has adopted a low profile and seems resigned to playing a secondary role. It is quite different with the so-called Asian Tigers – not just Japan, but also South Korea and above all China.

Investment in fundamental research has been growing year on year in China. With percentage increases we Europeans can only dream of, between 2000 and 2010 these investments were doubled, and today China spends more on research and development than is spent by the whole of

Europe. It has also launched an ambitious programme of space exploration which includes an orbiting science station and a lunar exploration mission. And every year it opens dozens of new universities and important research facilities.

The leadership in China has shown that it has understood that investment in basic science will allow the country to join the global technological elite. But their project is even more ambitious: they are not content to merely participate, they have decided to aspire to being the protagonists in activities which they consider will have a strategic importance for a superpower seeking to lead the rest of the world.

It is no coincidence that for physics the Asian giant is pushing ahead with the CEPC (circular electron–positron collider), a project which is similar to our own FCC: a ring of 50–70 kilometres that would host a *Higgs factory,* an electron–positron collider capable of 240 GeV, before moving to a proton accelerator capable of producing collisions with centre-of-mass energies of 50–70 TeV.

The machine may be built in the Qinhuangdao area, a hilly region near the coast, 300 kilometres from Beijing, known as the 'Tuscany of China'. To excavate a tunnel tens of kilometres long costs much less in China than doing something equivalent in Europe, and what's more the Chinese seem inclined to cover a good deal of the cost.

Basically, the FCC proposal, which comes at a time of crises and divisions throughout Europe, could offer the right opportunity to begin to rethink things on a larger scale.

If our continent aims to have a decisive role in the development of innovation and knowledge, and to resist all attempts to usurp its leadership in strategic sectors such as fundamental physics, FCC represents a great opportunity for it to confirm its centrality.

And so it is that the study of the origins of our universe – some 13.8 billion years ago – cannot fail to be entangled with the scientific technologies and perhaps even the politics of today.

Day Three: The Birth of the Immortals

The traumatic event that separated forever the weak force from its electromagnetic counterpart has just taken place, and apparently nothing has changed. The electroweak vacuum that has installed itself everywhere is invisible and cannot even be touched. But the components of that chaotic system feel it, a frenzy unleashed by pointlike objects that swirl everywhere.

The newcomer differentiates the behaviour of every single component, assigns roles, defines functions. It is as if, in this disorderly and indistinct system, an internal order had suddenly been established, albeit an invisible one – an order that will soon bring irreversible transformations. The apparent anarchy that dominates the multiple interactions now conceals a fine web of hierarchies and organisation. From this moment on, the changes will be profound. An imminent sequence of events will cause some elementary components to condense into ever more stable forms of organisation. These are the antecedents of a persistent material world, the small bricks are being cemented into the foundation of that great edifice, and soon we will be able to identify those elements that are familiar to us.

The universe has reached a dimension of 100 billion kilometres, and yet its expansion is unstoppable. Its temperature, despite its tendency to cool rapidly, can still be measured in thousands of billions of degrees. In the spasmodic agitation of its components we can begin to discern differences of behaviour and some regularity. In a few moments, with the decrease in temperature, the lightest of the quarks will freeze into a very particular state. A complex and ingenious system, a bound state of quarks and gluons which occupies a discrete part of the void; a very accommodating home with abundant room, a decidedly comfortable residence for three quarks and a certain number of gluons – a fairground for elementary components free to race after each other and to stick to each other, surrounded by virtual particles that bind them into a chaotic and vortex-like embrace. The environment is so well designed that it will last forever. The first protons emerge, the basic constituents of any more complex material, so solid and well organised as to be considered virtually immortal. Many other forms of organisation of matter will be unstable and will be converted into something else, perhaps after a fraction of a second or after millions of years. Things will be different for the proton, the average lifetime of which will be so immense that compared to it the 13.8 billion years that the universe has existed will appear to be an event of insignificant duration.

Everything is still incandescent, but soon the entire universe will come under the dominion of the Cold Genius, whose reign will not be short-lived, as it is in Henry Purcell's *King Arthur*. The great baroque composer will awaken him from his glacial sepulchre beneath the cloak of perennial snows, through the action of a primitive force that does not exist in the universe. The frozen environment that surrounds us, however, knows no spring; Persephone, the daughter of Demeter abducted by the king of the Underworld, has eaten all the pomegranate seeds and will no longer be able to climb back to the surface.

In this utterly inhospitable place, nothing is better equipped to survive than protons. They will constitute a kind of primordial cell of notes that will be used to compose the most complex symphonies. Combined in an infinite number of variations, they will give rise to the most unusual variants as well as the most reassuring reiterations, starting at a point of imminent novelty from which a sequence of other transformations will take their cue.

The specific mass that the electrons have assumed due to interaction with the electroweak void will allow them to orbit around the first protons in a stable fashion, so that they can form atoms and molecules. In this way the huge gaseous nebulae will be generated, from which the first stars and then galaxies will be born, as well as planets and solar systems – and the first living organisms that will gradually become ever more complex all the way down to our own era. That sequence of wonderful sounds is about to begin. Keep listening.

The Most Perfect Liquid

From the moment of the Big Bang, only a microsecond (10^{-6} or a millionth of a second) has passed, the temperature is more than 10,000 billion degrees, and the entire universe bubbles with a strange material. It resembles a kind of plasma, another word of Greek origin, indicating a gelatinous substance; something, precisely, that will be mouldable. It is the term used, for instance, for ionised gas, which is to say gas that has reached a temperature so high as to strip all its atoms of their electrons so that the medium, which remains electrically neutral, actually ends up composed of free particles of an opposite charge. The plasma that occupies the primordial universe is not composed of ions and electrons, but is made up of every kind of particle moving at relativistic speeds, above all of quarks and gluons. At such temperatures the strong force is actually too weak. Its coupling constant will gradually increase as the universe cools, but for now it is not yet able to contain the kinetic energy needed to create bound states.

The resulting plasma of quarks and gluons is a gas resembling an ideal liquid. Its components slide over each other without experiencing any resistance whatsoever, basically incapable of interacting with each other. It is a perfect liquid, with practically no viscosity; an ideal superfluid which flows everywhere easily and is capable of penetrating every interstitial space that has been left empty. This sort of thin soup – intangible, extremely hot and possessed of unusual properties – has been studied comprehensively since it became possible to create it in laboratory conditions. The findings are relatively recent and are based on the use of powerful machines equipped to bring heavy ions into collision with each other.

The most common accelerators use pointlike particles such as electrons, or at most protons, composite systems made up of a handful of quarks and gluons. In this case too the most energetic collisions occur between objects that are basically pointlike: their elementary constituents, namely quark or gluon pairs, collide head-on, while the rest of the proton shatters.

With special arrangements, it is possible to inject, put

into circulation and accelerate in the same machine objects that are much more cumbersome and complex, such as heavy ions. In effect we are dealing with the nuclei of ionised atoms that have been stripped of their orbital electrons, either completely or in part. Being charged, they may be injected into accelerators, acquire energy and enter into collision with other beams. Being heavier and more complex, their collisions are much more spectacular – real fireworks from which tens of thousands of particles emerge.

Let's consider the collisions between ions of lead that are produced in the LHC. In this case very heavy nuclei, consisting of more than two hundred protons and neutrons, are accelerated and made to collide, all brought to truly enormous energies.

An ultra-relativistic nucleus resembles a kind of thin, compact disc. Relativity squashes it in its direction of movement and the quarks and gluons of which it is composed, with their mass increasing with speed, cause a rapid increase in local density of nuclear matter. When two discs belonging to two different beams crash head-on, so to speak, it is as if there were a hundred overlapping individual collisions. At the heart of the collision a local temperature develops that is so high that quarks and gluons are seen, for a fraction of a second, fusing together to form a little droplet of that primeval fluid, the quark and gluon plasma.

The energy generated by most modern accelerators is so high that it can produce in the laboratory a miniature version of the Big Bang. The infinitesimal space in which this phenomenon occurs expands rapidly due to the extreme temperature, and in an instant the fluid is divested of its characteristic features, giving rise to streams of known particles. But the properties of these secondary products, emitted from the centre of the collision, allow us to trace the distinctive characteristics of the original superfluid.

Protons Are Forever

After a few microseconds, as the temperature cools, we go beyond the critical temperature that makes it possible for the plasma of quarks and gluons to survive. At this stage the universe is filled with a great abundance of photons, with quarks and leptons romping about everywhere together with gluons, while W and Z, which have become massive, have only a limited range.

As the universe cools, the interaction brought about by the gluons gets even stronger, every gluon ends up clinging to some quarks and disappears from view, and matter begins to aggregate into heavy states generically called *hadrons* (from the Greek for 'strong', as they are formed by quarks and subject to the strong force). The first attempts to produce stable matter fail: quark and antiquark pairs held together by gluons are born, but the bond is short-lived because it is unstable and readily broken. Everything takes a turn for the better when apparently more complex systems are formed, each with three quarks.

The new configuration is immediately more promising. The trio of quarks, held together by gluons that flitter between them, sticking to one after the other, looks like a system that's made to last. In reality, when heavier quarks are involved things do not turn out so well. For a brief moment it seems as if everything is fine, but then they too begin to show signs of instability before immediately, as the temperature falls, producing tiny fireworks as they disintegrate.

The real surprise occurs when the trio of quarks in question happens to be lighter. The first family contains the quarks *up* (*u*) and *down* (*d*) – the lightest and least perceptible, the ones that have the weakest interaction with the scalar Higgs field and are only heavier than the extremely light leptons. The gigantic *top*, thousands of times more massive, tries but fails to put together something that might be more stable. The little ones, however, manage to do precisely what their bulkier cousins have inevitably failed to achieve.

The architecture that emerges has the simplicity of brilliant ideas, like the three-legged table that always finds its equilibrium and never falls over. Two *up* quarks with a charge of +2/3, together with a *down* with a charge of −1/3, constitute a system with a net positive charge of +1 that will be called a *proton*.

The newcomer is a kind of archetype of stability, an ideal design, built to last. The group of three rotating quarks, entangled in the molasses of the strong force carried by the gluons, makes it into a kind of impenetrable fortress. Despite the lightness of its elementary components, it has a considerable mass, almost 1 GeV, dominated by the energy of the *strong force* field that holds it together. The three light quarks are connected by an enormous binding energy, far greater than their mass. It's the *strong glue* that keeps it together, the hidden secret of the proton's mass which gives it a legendary stability.

In an ever-cooling universe reaching energies lower than the energy that binds it together, it will become increasingly difficult to succeed in fragmenting protons. It will happen again when they are accelerated at ultra-relativistic speeds in stellar catastrophes and they will wander in the form of high-energy cosmic rays. The moment they collide with other bodies they will show the same disintegration reactions that humans will succeed in reproducing in their particle accelerators. But these will be rare phenomena. In the vast majority of cases the three light quarks, immersed in their ocean of sticky gluons, will remain becalmed and sheltered from the changes taking place in a universe that will evolve for billions of years.

Very complex experiments have sought to quantify the extent to which the proton is stable, that is, within what limits it may be said that the particle is *immortal*. The results are astonishing.

If a proton disintegrates into other, lighter particles, even through a very rare decay, its lifetime can be measured. It would be enough to identify one of these processes and the problem would be solved. Since we tend to expect these decays to be very rare, and it is impossible to conduct experiments that last for centuries, the only available option is to keep a truly frightening number of protons under control for a reasonable amount of time, such as perhaps a year or two.

In the Super-Kamiokande experiment in Japan, special sensors capable of registering the feeblest of disintegrations, are located in a huge container filled with fifty thousand tons of ultra-pure water. To avoid any possible false alarm, the smallest residual impurities are eliminated, and the whole thing is installed in a great cavern in the depths of a mine. In this way the experiment is less sensitive to the disturbances caused by cosmic rays, which could produce similar signals to those being sought.

Thus far, having not yet observed any decay, it has only been possible to establish the lower limits of the lifetime of a proton, which turned out to be greater than 10^{34} years. In short, within the limits of the experiment, its life is eternal. It is enough to be reminded that the age of the universe is slightly greater than 10^{10} years. To adapt a phrase originally coined to advertise jewellery, 'protons are forever'. If it is true, however, that as far as longevity is concerned there is no competition between a diamond and a proton, it remains doubtful as to whether a gift of a small jet of hydrogen will do as a replacement for a sparkling ring.

The interest in looking for those very rare processes in which a proton might decay into other, lighter particles is increased by its connection to the experimental validation of a Grand Unified Theory (GUT). That the three fundamental interactions converge in a single force for energies that are sufficiently high is widely considered to be a very convincing hypothesis, and is supported by much experimental data. Since the unification would manifest itself at a scale of energy that is currently not achievable, it is not possible to observe the phenomenon directly, nor to study it in detail. Some theoretical models of GUT predict that, however infrequently it may happen, even the proton must decay. So at present the discovery of this process, which is so difficult to trace, might well afford us clear indications regarding the dynamics of Grand Unification.

It is already possible to anticipate that protons constitute the principal component of ordinary matter in the universe. The majority of the visible matter of galaxies takes the form of hydrogen plasma – hot ionised gas consisting of free protons and electrons. If the protons were unstable, the plasma would dissolve like mist under the rays of the Sun. But this does not happen. The protons – be they free to meander in space or strictly attached to atomic nuclei – seem to be to all intents and purposes immortal. Like the warriors in *Highlander*, a movie from the 1980s starring Sean Connery and Christopher Lambert, the protons have experienced the vicissitudes of the universe since time immemorial, and nothing seems to worry them about the future.

Light yet Indispensable

Two of the ingredients needed to construct the stable matter that we are all familiar with are still missing. The first is the neutral version of the proton: the neutron, a close family member which resembles it in many respects. It too is made up of a trio of light quarks, except that it contains two *down* quarks (each with a charge of −1/3) and one *up* (with a charge of +2/3). The result is an object that is also massive but lacking electrical charge. The mass ends up being similar to that of the proton, approximately 1 GeV, dominated also in this case by the binding energy of the gluon field that holds it together. But the fact that it is neutral creates a very small but significant difference. The neutron is slightly heavier than the proton, a barely discernible 1.3 MeV (mega electron-volts), equivalent to an increase of 0.14% – but the difference will turn out to be crucial.

Having a mass only slightly greater, it may decay into a proton, and in order to comply with the laws of conservation of energy, into an electron that must necessarily be accompanied by a neutrino. It is a typical weak decay with an emission of electrons, similar to that which had intrigued Enrico Fermi. This decay does not occur if the neutrons are packed within the nuclei. In the field of the strong force that holds the nuclei together, the neutron cannot decay – but if it cannot rely on this protective shield, it becomes unstable and disintegrates after a few minutes. In due course we will see the important role it played in the formation of the first nuclei.

Protons and neutrons form continuously, together with their related antiparticles. When the two opposites meet, they immediately annihilate each other to produce photons, but the environment is so hot that it continues to extract from the void particle–antiparticle pairs to replace those that have just disappeared. The process is repeated everywhere, continuously, for as long as the temperature permits. In this fleeting cycle of extremely rapid birth and death, the small initial asymmetry between matter and antimatter is amplified. Slowly but inexorably, that infinitesimal difference in population results in all the antiprotons and all the antineutrinos disappearing from subsequent generations. This is how the universe moves towards being made up only of matter.

Then the temperature drops below the minimum value that makes it possible to extract pairs of protons or neutrons from the void, and the process comes to a halt, signalling the end of the hadron era. There will still be enough energy to produce electron–positron pairs that will begin to populate the universe, repeating a story similar to that of the hadrons.

Unlike protons and neutrons, electrons are extremely light. In fact they weigh almost two thousand times less than the quark triplets they would like to accompany. They are not composite objects, and there are no charged particles lighter than they are. Combining conservation of energy – whereby an object can only decay into a lighter particle – with that of charge – whereby an electron may not decay into a neutral particle – we conclude that electrons must ultimately be stable.

After a few instants have elapsed since the Big Bang, the universe fills with the lightest of the charged particles. Now it contains all of the essential ingredients that allow it to form stable matter. But a little more patience is still required at this point.

The Shyest and Kindest Are the First to Leave

Since the universe began to fill with protons and neutrons, the population of neutrinos also increased. They are the lightest of the leptons, and have such easily overlooked masses that they had deceived us until a few years ago. Only recently was it discovered that their mass is very slightly different from zero, even though we have not yet managed to measure it precisely. They are leptons, so they do not experience the strong force; and they are neutral, so they are indifferent to the electromagnetic interaction. The only force they obey is the weak force. This makes them less invasive; in fact it makes them exquisitely gentle. Neutrinos are very reserved particles that move with great delicacy, to the extent that they are able to cross vast quantities of matter without being noticed, and without producing the slightest disturbance. And yet they play a crucial role in establishing the equilibrium that will determine the material composition of the universe.

As we have seen, neutrons are slightly heavier than protons; their 0.14% difference is insignificant: it's as if between two individuals, each weighing 80 kilos, one were to give prominence to a difference between them of 100 grams. That said, if protons and neutrons must assume a thermal equilibrium with each other, they must each absorb half the energy. Owing to the difference in mass, the population of neutrons will be slightly smaller than that of protons. While the temperature remains huge, this small detail may be dismissed. But as the thermal energy that needs to be distributed diminishes, the difference grows in significance. And what is responsible for reducing the population of neutrons and increasing that of protons? Reactions that make the neutron disappear, such as weak decay, which transforms it into a proton, an electron and a neutrino, and other reactions with similar effects. The conclusion to be drawn from this is that the gas of neutrinos that takes part in these processes ends up sharing the same temperature as the population of photons and of the hadronic matter with which it interacts.

This dynamic process is interrupted the instant the universe is 1 second old. At this point the temperature has declined so much that to maintain thermal equilibrium there are now six protons for every neutron, and the situation is about to come to a head. The temperature now drops so rapidly that the neutrinos no longer manage to maintain the correct reaction rate to distribute the thermal energy between protons and neutrons. Scarcely a moment ago the different kinds of particles were kept in equilibrium. Now there is a Caporetto-style rout: the battle having been irretrievably lost, the neutrinos abandon the field. An enormous population of delicate and peace-loving particles detaches itself from primeval matter and begins to wander aimlessly, carrying with it only the residual memory of the temperature shared with their partners a mere instant before separation occurred.

From this moment, in a universe too rarefied now to hold them, the neutrinos escape from the grasp of aggregated matter and will no longer be able to repeat that primal embrace. They will wander indefinitely, for billions of years, assisting in the formation of stars and galaxies, enormous distributions of matter that they will continue to traverse, hardly noticed, with their signature delicacy.

Their evolutionary history will be different, but the memory of the golden age – that extremely hot and magical era in which they too played hide-and-seek with matter and coupled freely with a multitude of particles – will remain indelibly encoded in their temperature. Today, after 13.8 billion years, the extremely ancient cosmological neutrinos – as they are called in order to distinguish them from the very recent ones produced by stars – still continue to wander everywhere. According to our calculations, every square metre of the universe should contain six hundred, which seems like quite a good number, but neutrinos interact so weakly with matter that so far nobody has managed to gather evidence proving their existence. And yet we are certain that they still surround us. We also have an idea of their probable temperature, which due to the expansion of the universe should be around 1.95 degrees kelvin.

For the moment the search for signals from cosmological neutrinos has not yielded particularly significant results. Up to now, only vestiges of their presence have been found. The day that some new technique will reveal them to us we will be able to study all the properties of the cosmic background of neutrinos, hypothesised by all the models of the Big Bang. This sea of shy and benevolent particles by which we are still surrounded contains secrets that will prove decisive for understanding what really happened when the universe blew out the little candle of its first second of life.

They Will Form the Heart of Stars

At the end of the first minute there are already seven protons for every neutron, and the density of energy has lowered to the point at which they can begin to aggregate among themselves and form the nuclei of the lighter elements.

It is a fundamental moment, because the density and temperature of the universe now resemble those of the stars. Protons and neutrons, involved in high-energy collisions, may react to form bound states through the strong force. When a proton couples with a neutron, it becomes the nucleus of deuterium; if two deuterium nuclei fuse, the first nuclei of helium are born. This light element, the nucleus of which is formed by two protons and two neutrons, gets its name from the Greek god of the Sun – and in effect all the hydrogen that feeds the immense nuclear furnace of the stars ends up by becoming helium.

To form its nucleus the two nuclei of deuterium need to fuse together, a process that occurs very easily. The four-part nucleus is very stable because it involves a very high binding energy for each of the components. All of the free residual neutrons will be involved in these quadrilles and vanish from the scene. That is why helium nuclei will account for around

24% of the total mass. The rest will be made up of protons that will remain 'single', ready to be transformed into atoms of hydrogen as soon as circumstances will allow. Here and there, in traces, nuclei of a slightly heavier kind will appear, such as those of lithium and beryllium.

There will only be three minutes for the formation of all the originary nuclei in the universe. At the end of the three minutes, the temperature and density will no longer be high enough to sustain nuclear reactions. And this will be a good outcome, since if the process had continued for a prolonged period the universe would have consumed a vast quantity of free protons in order to construct heavier nuclei. If it had lasted for just ten minutes, almost all of the hydrogen would have disappeared.

The abundance of helium in the universe is a further confirmation of the Big Bang theory. This element is also produced at the heart of stars, but without primordial helium things would not add up. Not even with all the stars that we know to exist in the universe burning hydrogen for 14 billion years would it be possible to produce the abundance of helium that has been measured.

The nuclei created at this time will not undergo any further modification for billions of years, and even today they make up the majority of nuclei existing in the universe. To these will be added, much later, the nuclei of the heavy elements of the periodic table, which will be born in the immense nuclear furnaces of the most massive stars.

Theoretical calculations have estimated that if the difference between proton and neutron mass had been only a little greater, the outcome would have been disastrous. A small detail, the beat of a butterfly's wings, and a series of catastrophes would have followed. The difference in mass would have significantly altered the proportion of protons to neutrons, and we would have a great deal more helium and much less hydrogen. In short, there would not have been enough hydrogen to unleash the nuclear reactions in the first stars. Everything would have remained shrouded, for all time, in deepest darkness; the universe would have remained an immense, gloomy, pitch-black space; without stars there would have been none of the heavy elements and the primary material for rocky planets would have been missing. We would not have the conditions for fostering the elementary signs of life, and therefore that being who one day would be able to contemplate their grandeur would also never have existed.

Fortunately none of this happened in our universe. The tightrope walker got safely across: it seemed that at any moment he might fall to one side or the other, while the public suppressed a tremendous gasp at the imminent tragedy – and yet with skill and even elegance he always managed to regain his balance, finishing the feat amidst thunderous applause.

It will take a great deal of time before energy has decreased sufficiently to permit the formation of the first atoms of hydrogen. We will have to wait for the moment when the temperature of the universe is low enough not to break the electromagnetic bond that will allow electrons to orbit around the protons of the nucleus. But we have made very important progress by the end of the third day – and only three minutes have passed since the beginning of this extraordinary adventure.

Day Four: And Then, at Last, There Was Light

The first minutes having passed, there is a brutal and totally unexpected change of rhythm. The sequence of spasmodic transformations that the universe has gone through abruptly stops, and everything slows down to such an extent that it is almost lost altogether, entering a process of such laboured slowness as to appear potentially exhausting. We have barely recovered from the *crescendo in prestissimo* with which the symphony began, and are waiting for a passage of a more regular and reassuring tempo, when everything plummets into a *larghissimo* that seems to lead nowhere.

The processes have now become infinitely slower, and eras disproportionately prolonged. Waiting for the most significant developments will require a great deal of patience. After the formation of the nuclei of the lighter elements, nothing major will happen for hundreds of thousands of years. Except, that is, that everything continues to be subject to expansion and cooling.

For a period that seems endless, the universe is filled with a dark fog. This is an opaque world, made up of elementary particles and nuclei, all intermingled and immersed in a sea of photons and electrons. Unknown particles of dark matter clandestinely contribute to this apparently perpetual sarabande. There is no structure, no hierarchy, no organisation. Absolutely nothing.

Not a single ray of light manages to penetrate this dark and disturbing plasma. Electrons and photons race after each other, playing a game of tag. Continuously absorbed and then immediately emitted by the dense gas of electrons that infiltrates everything, the photons cannot escape from this suffocating embrace.

This opaque kingdom will last for hundreds of thousands of years. There is no other reign of darkness to compare with it: not even the most imaginative science fiction can rival the hostile atmosphere of this boundlessly dark and dispiriting environment.

The key to its transformation comes, as usual, from a change in temperature. The temperature has been falling, gradually and irrevocably, as the universe continues to expand. Everything changes as it approaches 3,000 degrees. This is approximately half the temperature that is measured on the surface of the Sun – and at this point the opaque fog begins to disperse. With the falling temperature, the kinetic energy of electrons diminishes, and they can no longer break the electromagnetic attraction that binds them to protons. Electromagnetic attraction prevails: a myriad of feral electrons wandering freely now become domesticated by the electromagnetic field. They will no longer enjoy their freedom to roam, but be obliged instead to orbit in a stable fashion around a charged nucleus.

The first atoms are formed, above all those made up of hydrogen and helium. They emerge everywhere, the plasma breaking down into an immense quantity of gas that implacably absorbs all the nuclei and the entire population of electrons. Matter begins to assume a neutral and stable form. In a breakthrough moment, the atoms make it possible to build ever more complex structures that will lead in turn to further transformations.

While electrons become resigned to the end of their freedom, trapped as they are in the comfortable shell of atomic orbit, for photons it means the end of prolonged slavery. Suddenly emancipated from their ties to matter, they can now run free – and celebrate the ability to do so by carrying light into every corner of the universe. At a stroke the universe becomes glaringly lit and transparent.

From now onwards, photons will shoot everywhere, bouncing off everything. With the passage of time they will become ever less energetic, and their frequency will diminish – an unequivocal sign of weakness. Submerged in an ever colder thermal bath, they will continue to oscillate, but ever more feebly. Even so, they will be able to carry with them the indelible memory of the epoch in which the world was dominated by radiation, and matter organised into atoms did not yet exist.

In short, there was light. Just as the Bible has it, except that it did not happen suddenly and it was by no means easily achieved. At the end of the fourth 'day', 380,000 years have elapsed.

A World without Light and Filled with Dark Entities

After a period of just a few minutes in which nuclei are formed, nothing significant happens for thousands of years. The expansion and 'cooling' down of the universe continues incessantly: it is soon more than a thousand light years in extent, though its temperature can still be measured in millions of degrees. It is an enormous object that is both incredibly hot and dark. A hellish world, without light and filled with dark entities.

A kind of opaque and impalpable fog fills and envelops it. An aerosol of electrons, photons, and other elementary particles surrounds the protons, nuclei of helium, and the rare light elements that have formed thus far.

The temperature is still too high to permit matter to aggregate as a result of electromagnetic attraction. Protons and helium nuclei, positively charged, try to attach themselves to the electrons that are flying around – but fail in the attempt. Thermal turbulence makes the electrons so supercharged that even if a link were to form, it would not last for more than a fraction of a second. The force of attraction is too weak to compete with the frenzied kinetic energy that carries them away into the distance. Everything will have to wait a long while for a drastic fall in temperature, before being able to celebrate the great triumph of electromagnetic bonds.

All the material particles are immersed in a bath of photons that share the temperature of the system, but there is no trace of light. The density of this strange fog that envelops the universe is so high that every photon is constantly absorbed and immediately emitted.

The embrace between photons and matter, especially with electrons, is oppressive, denying any freedom: the average leeway they have in which to move is infinitesimal. Every time they are emitted by an electron that enters into a collision, or is accelerated, they hopefully embark on what they long to be extended distances and prolonged races towards the infinite, only to find themselves immediately swallowed up by something else, without time to reflect on their melancholy destiny before the endless cycle of emission and absorption has begun again.

In the darkness of this strange world, forms of matter are hidden that are even more mysterious. We've only hinted at them so far because we're not sure exactly what they are. It is therefore quite difficult to assign to them a precise location in the sequence of events that has brought us this far. But we can definitely say that in the cosmic dark ages there is already present in the universe a significant amount of dark matter.

The hypothesis that the universe contains great quantities of non-luminous matter was first advanced in 1933 by Fritz Zwicky, a Swiss astrophysicist possessed of great ingenuity and a mischievous sense of humour. It's said that when other scientists voiced any scepticism concerning his theories, he would insult them by calling them 'spherical bastards'. He would then explain to his bewildered interlocutor that they were bastards from whatever angle they were looked at.

Working on the Coma cluster, an agglomeration that we now know contains more than a thousand galaxies, Zwicky noticed that there was something wrong with the speed of those that were nearest to the edge of the cluster. Their motion could not be explained with the distribution of visible mass obtained from light. The effects of gravity were not sufficient to explain speeds such as those of the outer galaxies. Everything seemed to be behaving as if the volume of the cluster concealed a much greater amount of matter. Zwicky calculated that a vastly greater mass was necessary, which remained hidden in the darkness of the cosmos and which he called 'dark matter' because it emitted no light. For a good while this theory of his attracted virulent criticism, and the number of 'spherical bastards' showed no sign of decreasing.

The status quo was overturned by the work of Vera Rubin, an American astronomer and the heir of Henrietta Swan Leavitt who had invented the Cepheid method for measuring large distances. Rubin was one of the very few astronomers who, even in the 1960s, had access to the largest telescopes. It is worth noting that when she started working at the Mount Palomar telescope, she had to personally arrange for the provision of a women's bathroom because the architects of the most modern observatory in the world had not conceived of the possibility that a female astronomer might work there.

Very systematically, Rubin measured the rotation speed of the stars within spiral galaxies. She started with Andromeda, and she confirmed that the outermost material orbited at speeds comparable with those of the inner stars: a result opposite to the expected one, given that the gravitational attraction produced by the only luminous material would have involved a much lesser speed. Comparable observations were made with regard to the movement of whole galaxies within the cluster, and the conclusion was inevitable: the apparently eccentric Zwicky was right. Rubin's calculation demonstrated that dark matter had to be at least five times more abundant than luminous matter. The spiral galaxies had to be immersed in a gigantic shroud of completely unknown matter, without which they would have long since disintegrated.

In the second half of the twentieth century, evidence of the presence of dark matter has been recorded by numerous experiments. Diverse methods of inquiry have all arrived at the same result: there will be indirect evidence of dark matter when we are able to measure the rotational speed of the immense cloud of hydrogen that surrounds many galaxies, and also when observations made with gravitational lenses have increased. Even this phenomenon had been foreseen by Zwicky, who described it as a necessary consequence of general relativity. The inspired Swiss astronomer was the first to understand that a strong concentration of masses might have warped space-time in a way that created the same optical effects as a lens. The luminous rays that had crossed the warped zone could produce, by being diverted, the most incredible artefacts. The same star or the same galaxy could appear two, three or even four times in the image captured by the telescope.

These ghosts, these double-exposure images that could lead us to suspect that someone has had a few too many drinks and was seeing double, will in fact turn out to be innovative instruments of measurement that will allow us to see forms of matter that would otherwise remain invisible.

Despite ever more convincing experimental evidence and the fact that no one had ever questioned the relevance of her discovery, the Nobel committee failed to award to Vera Rubin the prize she so clearly deserved. No one knows why.

Today we believe that approximately one quarter of the universe is made up of dark matter, but nobody knows yet what it really is.

It has been suggested that it could be a gas made up of neutrinos, but given their lightness it would be difficult to account for the gravitational effects that we observe. If the theories of supersymmetry were correct, there would be entire families of new particles, very heavy and with strange names, that could explain dark matter. But given that no supersymmetric particle has been discovered yet, the hypothesis that the halo surrounding the galaxies is made up of *gravitinos* or *neutralinos* appears to be altogether arbitrary at present.

The hunt for every kind of heavy and weakly interacting particle that could explain this enigma is still very much ongoing. More sophisticated experiments are being organised in underground laboratories; equipment is sent into orbit around the Earth; the most powerful accelerators are engaged: all, so far, to no avail.

There are those who think that instead of looking for heavy objects, we should concentrate our attention on neutral particles that are in fact ultra-light, known as *axions*. The name was coined by Nobel laureate Frank Wilczek, who borrowed it from a well-known detergent of the 1950s, presumably convinced that this new particle would definitely clear everything up. Axions are in effect small, evanescent, extremely light corpuscles, hypothesised to explain small anomalies in the decay of known particles. They would be capable of interacting with ordinary matter almost exclusively thanks to gravity. But at the moment there are no signs of confirmation of this hypothesis, and so the hunt continues.

Whatever the solution to the puzzle might turn out to be, dark matter certainly came onto the playing field in one of the preceding phases, perhaps immediately after the inflationary phase. Cooling down like everything else, it started to show extremely small temperature differences in its energy distribution, having initially been perfectly homogeneous.

These differences are born from elementary quantum fluctuations magnified by the inflationary expansion and due to the interaction with the turbulent sea of photons that is everywhere in a constant state of agitation.

Now, during the era of opacity, we imagine it as a sort of fine net – a black, tenuous spiderweb that mixes and binds everything together. Its spatial distribution does not for now play a role relevant to the dynamics of this dark plasma, but very soon it will trigger a concentration mechanism that will lead matter to become denser wherever there are minute energy fluctuations. The denser knots of this thin spiderweb will provide the weft on which our material world will begin to thicken. There the first stars will be born, and the seeds of the galaxies will bloom.

The Hour of Matter Strikes

The gloomy reign of opacity was so prolonged that it almost seemed as if nothing could disturb its equilibrium.

But when the temperature dropped below 3,000 degrees, something irreparable happened. This value marked a border that once crossed would initiate a sequence of irreversible, interrelated phenomena. Hundreds of thousands of years have passed since the Big Bang, and up until this point the components of matter have remained completely immersed in the sea of photons of cosmic radiation, sharing its temperature. Thermal equilibrium is guaranteed by the continuous interactions between the two, made frenetic by high density. But with the expansion we get to the point at which everything changes at a stroke.

Everything has to do with a difference in behaviour between

radiation and matter that it is worth taking the trouble to underline. The expansion of the universe increases volume by the cube of its radius: like an inflating balloon, doubling the radius results in an eight times greater volume. The density of matter and energy decreases with the growth in volume, that is to say, in inverse proportion to the cube of the radius. For the photons of radiation, instead, a further mechanism comes into play that ultimately reduces their density. With space stretching, their wavelength increases and hence their energy decreases. In short, the density of energy due to radiation diminishes more rapidly than the density of energy due to matter. With a doubling of the radius, the energy of radiation becomes sixteen times smaller, whereas that due to matter is reduced only eightfold.

In the long run, the equilibrium breaks down catastrophically. This happens 380,000 years after the Big Bang. At that moment radiation separates from matter, and after that separation their respective destinies completely diverge. The density of the photons will thin out to the point where interactions with electrons and protons will be less and less frequent and the thermal equilibrium will be broken. So begins a long decline that will see the radiation that until this moment had dominated the world assume an increasingly minor significance, until it becomes an irrelevant component of the total mass of the universe.

Very soon the temperature will fall to the point at which the potential energy of the electromagnetic linkage between electrons and protons will outstrip the kinetic energy of thermal turbulence. Electrons will then be able to bond stably with protons, and the first atoms will be born, notably hydrogen and helium, then lithium, beryllium, and some other light elements. Freed of constant interaction with photons, the atoms will find their own stability.

From this new order of things, neutral matter emerges that will naturally interact less with radiation. An immense and rarefied cloud of hydrogen and helium will occupy the whole universe, and its evolution will determine the rest of our story. After millennia during which radiation dominated the universe, this traumatic separation signals the beginning of the era of matter. The new epoch will lead to the formation of galaxies, stars and planets – and to the development of material forms of a more complex kind that will become living organisms. A new kind of dominion is established, a reign that will last for billions of years and that even today has no end in sight.

As for the photons, released finally from the ties that imprisoned them, freed from that seemingly inextricable embrace, they can at last travel anywhere freely. The sea of photons retreats from matter, but occupies every space left free by nascent atoms, bringing with it a distinct form of energy. The universe becomes transparent, allowing light to cross it from side to side. It is a glow different from the white light to which we are accustomed. Our eyes, if we can be allowed the absurdity of imagining ourselves there as witnesses of the new phenomenon, would see a kind of reddish flash: a hot light that goes beyond the dark red which marks the upper limit of the wavelengths visible to humans. Curiously, it is a lot like the light emitted when using the TV remote to change channels. But there can be no doubt: it is light. The universe is transparent, penetrated by light.

The Secret Messages in the Wall

Twice every year at Judaism's most sacred site in Jerusalem, the Western Wall is cleared of all the small notes that according to ancient tradition have been inserted by the faithful into the gaps between stones. Employing small tools, a group of helpers extract with great care and delicacy the small sheets of paper stuck in the narrow cracks, thus creating space for those that will replace them in the coming months. The folded notes are not thrown away, but buried in the Jewish Cemetery on the Mount of Olives, a small hill not far from the city's historic centre.

The Western Wall is part of a city wall built by Herod the Great, king of Judea, during the Roman occupation. Work on this structure started in 19 BC and was completed in AD 64 with the aim of consolidating the hill on which the Second Temple and Judaism's most sacred space arose. In AD 70 Titus's troops profaned the sacred place and razed the temple to the ground. It was never rebuilt. It was the end of a world, an apocalypse. The only surviving trace of the original construction was the retaining wall erected by Herod – and this is the structure venerated ever since by adherents to the Jewish faith, as a place of prayer as well as a site of remembrance for one of the most traumatic and painful events in the history of that faith. For centuries people have gone to the wall to weep and pray, recalling the terrible misfortune that led to the Diaspora. The Western Wall is also known as the Wailing Wall because of the way that pilgrims expressed the emotion they experienced on touching the ancient stones, as they placed the palms of their hands and their foreheads against them to pray.

Beginning in the medieval period it became common

practice also for pilgrims to leave signs of their visit: incisions, graffiti, even the imprint of hands that had been dipped in plaster. With the passage of time these habits were prohibited because they were in danger of irreparably damaging the ancient stones, and were replaced instead by the custom of leaving behind little pieces of paper in the gaps between them. This tradition continues to the present day, but by now the number of visitors has grown to such an extent that it is necessary to periodically remove them in order to make way for more. These pieces of paper contain prayers and requests for aid. They are highly personal invocations, often reflecting the tribulations and secrets of the families that leave them. The hopes and sorrows of generations of the faithful are hidden and accumulate in the crevices of the wall.

Something similar happens in another type of wall – albeit a much less material, definitely less tangible, and inconceivably older one. I mean the 'wall' of cosmic background radiation.

The light that separated from matter in that extremely distant period has for billions of years retained the memory of that traumatic event. The primordial photons, the first to taste the euphoria of liberty, are still all around us and fill the cosmos in every direction. With the passage of time, their temperature has plummeted from 3,000 to less than 3 degrees. In this time the universe has increased to a thousand times its former size, and the stretching of space-time has vastly increased their wavelength. Now they no longer oscillate on infrared frequencies; their song has acquired a much deeper tone, almost impossible to hear, ending up in the region of microwaves. Yes, it is practically equivalent to the radiation we use in our kitchens to defrost something. This in effect goes for the whole universe: unable to exchange energy with any other system, it behaves like a gigantic microwave oven, an enormous black body that follows the same laws.

The amazing thing is that in the sea of photons of cosmic background radiation the unmistakable traces of this era have remained impressed, like the fossils found inside certain types of rock. The last contact with matter, an instant before separation, has left definite traces that have been gradually attenuated but still allow us to gather from them precious information, enabling us to follow the path back to the epoch when matter and radiation went hand in hand – and even further back still.

It is the dream of every scientist to be able to see so far back in time as to witness at first hand, through a telescope, the birth of the universe: to watch the Big Bang itself. Using light, the photons of electromagnetic radiation, this dream soon proves to be impossible, because once you reach 380,000 years after the event you come up against a kind of wall, albeit one that may contain valuable information hidden in its interstices. By gathering and interpreting this information, scientists have managed to understand the secrets of the instant in which the predominance of matter began, and along with these secrets they have gathered data of inestimable value regarding everything that happened before, managing even to reach back as far as the instant of the first momentous transformation, to signs of cosmic inflation.

A Very Detailed Story

Cosmic microwave background (CMB) radiation is our most valuable source of information on the origins of the universe and the transformations it has gone through.

Ever since its discovery in 1964 by Penzias and Wilson, increasingly sophisticated experiments have yielded an impressive volume of results. CMB can be seen as a kind of mine, the seams of which – extremely rich deposits – have already supplied a vast amount of data. But there is still a considerable amount that needs to be excavated, and we know that there are hidden veins that have yet to be exploited, a real treasure trove of information.

Reconstructing the low-energy photons of which it is made, coming from all directions, it is possible to derive a picture of the heavens in their entirety, and to extract from it a remarkable mountain of information.

The first characteristic is the extreme homogeneity of temperature distribution. CMB has an ideal black-body spectrum, and radiation is so feeble that the temperature of the universe is 2.72 degrees above absolute zero. The hypothesis that the universe behaves like a vast, ideal, and perfectly isolated oven appears in fact to be correct. The first photons, which after separating from matter have continued for billions of years to cool down, still remember that they operated in thermal equilibrium with it for 380,000 years. The flow of radiation is uniform in all directions, but there are microscopic areas characterised by extremely small differences in temperature that show evidence of a very characteristic structure.

These irregularities or anisotropies in the distribution of temperature have been studied in great detail because they contain a wealth of information about what happened in the first instants of the life of the universe. They are similar to the small pieces of paper inserted into the cracks in the Western Wall; they tell each other secrets and ancestral stories. They show the imprint, a legacy of radiation, of the quantum fluctuations that used to ripple the surface of the tiny bubble that emerged from the void, before it was magnified out of all proportion by inflation. Those parts of space that were once infinitesimal have expanded to grotesque proportions and now cover an area that holds entire clusters of galaxies. In the psychedelic heavens reconstructed by the most recent experiments, such as those conducted by the Planck satellite that concluded its mission in 2013, it is possible to see the working of quantum mechanics on a galactic scale.

The old prejudice, according to which the theory proposed by Planck and Heisenberg would explain only the phenomena of the infinitesimally small, has been conclusively overcome on the basis of observational data. Cosmic background radiation provides a clear, easily readable map of the density of matter at the point of separation from photons. Every minor local temperature difference may be attributed to a difference in density of the materials at the instant the photons were subjected to diffusion, an instant before separating forever. This allows us to visualise the vast cosmic spiderweb around which the first seeds of galaxies were constructed.

By analysing in detail the distribution and dimensions of small irregularities, it is possible to recover precious information about the geometry of the universe.

A closed or open universe would deform in a characteristic manner the image of such distant objects, because the photons would run along convergent or divergent lines. From the dimensions and angular distribution of these inhomogeneities we obtain unequivocal confirmation that our universe is flat. This implies that the density of matter is very close to the critical density. Cosmic background radiation thus gives us further confirmation of the presence of dark matter and energy in proportions that today we can measure with precision. The most recent data indicates that the universe is made up of 68% dark energy, 27% dark matter, and only 5% ordinary matter.

By simulating the distorting effects on the picture due to the curving of space-time by dark matter, it is possible to draw a map showing its distribution. The gravitational lens effect allows us to obtain, from the CMB, a three-dimensional image of the distribution of dark matter in the universe. To know in detail how this fine cosmic spiderweb is shaped allows us to understand better the mechanisms that lead to the formation of the first stars and the first galaxies.

Quantitative analysis of the distribution of fluctuations of primeval temperatures in cosmic background radiation provides us with one of the most solid confirmations of the existence of inflation. Looking ahead it is expected that soon we will be able to obtain new and more complete results from the measurement of its polarisation.

The polarisation of radiation indicates whether electromagnetic waves vibrate in a preferential direction. It involves the same mechanism that has made Polaroid sunglasses so successful. The reflections of the Sun's rays on the surface of water, for instance, consist of polarised light, which is to say that the electromagnetic field of the rays oscillate only on the horizontal plane. If you use a vertical filter, a fine membrane that allows only waves that oscillate vertically to pass, the annoying reflections are absorbed. Polarising lenses are lenses in glass or plastic inside which there are these vertical filters that absorb the reverberations of light largely responsible for blinding glare or visual discomfort.

Cosmic background radiation ends up polarised through interaction with the material medium, and hence carries with it additional information on the history of the cosmos. This characteristic tells us something further about the contact between radiation and matter. Forms of linear polarisation can be combined with the density of matter, thus furnishing more details, for instance, on the distribution of dark matter at the moment of separation.

The most modern experiments have been successful in measuring this weak polarisation, obtaining important results in doing so. The polarisation most sought after, though as yet without success, is of a vortex type that would have been produced by the interaction of photons with primeval gravitational waves. It involves an even more subtle effect, an extremely weak polarisation that is also masked by similar phenomena produced by intergalactic dust. Finding it is a nightmare for experimental physicists.

If it proves possible to identify the signal left behind by the last encounter between photons and gravitational waves, it will further represent an unmistakable imprint of inflation. That strange polarisation which scientists have been trying to identify for decades could be the key to unlocking the chest that still contains many of the secrets of the inflationary phase. It would, for instance, make it possible to determine the scale of energy required to facilitate the initial fluctuations generated in the very first fractions of a second after the Big Bang.

To better understand inflation, scientists have other arrows in their quiver with which they might hit the target. To distinguish between the various different scalar fields that could have triggered it, consideration is ongoing as to how to study with still greater precision the large-scale structures of early galaxies. Their distribution should follow the traces of the minute fluctuations of the field of inflatons that thanks to inflationary expansion have remained imprinted in cosmic background radiation. It will be necessary to gather the most extensive sample of primordial galaxies possible, to observe in effect the most remote galaxies when they were undergoing formation – and this is what a new generation of experiments soon to be inaugurated in space are designed to achieve. With the aid of cosmological neutrinos and fossil gravitational waves that will sooner or later be identified, the secrets of inflation should soon be completely revealed. Always assuming, that is, that there is no additional surprise of some other scalar in LHC data.

We have now come to the end of the fourth day, 380,000 years have passed since the initiation of the Big Bang, and the universe is entering a very interesting phase: a sequence of transformations from which the first star will emerge. A part of matter is about to organise itself into a new dynamic and turbulent form that will illuminate the universe and turn it into a magnificent spectacle even for our very limited vision. From the enormous furnaces that will be ignited at the heart of stars, those heavy elements will be born that are destined to produce other, calmer, less turbulent forms of aggregation: the planets. Here they will be transformed into rocks, air, water, plants, animals and ourselves. If we can begin to accept the idea that we are literally children of the stars, we should also accept the fact that we are great-grandchildren of those quantum fluctuations that have expanded since inflation and without which the first stars would not have been able to form.

Day Five: The First Star Lights Up

The era of matter has just begun, and the rhythm of transformation increasingly slows. Up to this stage the weakest of the interactions, gravity, has occupied a somewhat marginal position. Now its presence starts to count, delicately and almost imperceptibly at first, but soon enough it will arrogantly command centre stage.

With the uncoupling of matter from radiation, things have acquired greater clarity. Radiation has been distributed uniformly throughout available space and the universe has become transparent with light. But the glow that marked the last metamorphosis has vanished since that expansion stretched the wavelength beyond the threshold of the visible. The universe is filled with radiation, and it is still very hot, but it is now plunged once again into total darkness.

Matter moves slowly, under the influence of gravity, and it stabilises into atoms which form an immense cloud of hydrogen and helium. Protected by the darkness, an enormous spiderweb of dark matter, already more abundant than ordinary matter, envelops the entire cosmos.

The little anomalies in its density, born of the quantum fluctuations which preceded inflation, have expanded disproportionately – and now something is beginning to happen around these areas. If we could go beyond the dark veil that hides everything, we would find ourselves present at a slow but inexorable growth in the density of gas. In these anomalous regions with their irregular contours a slightly higher than average density exists, and the gravitational force arising from it attracts other matter. In this way ever more significant agglomerations occur, and while this is happening the distribution of matter acquires an increasingly marked spherical symmetry.

The process is very slow and will take millions of years. But although the speed at which it advances is almost imperceptible, the pace of gravity is relentless: nothing will stand in the way of its dominance in the newly formed material universe.

Around the irregularities vast concentrations of gas begin to thicken; here and there it becomes possible to distinguish spherical bodies of truly enormous mass, at least a hundred times heavier than the Sun.

The force of gravity that develops from them is immensely strong: it compresses the gas, pushing it violently towards the centre of the system which gets hotter and ionises the hydrogen. The vast celestial body now consists of external layers of gas and an innermost core of extremely hot plasma. The relentless grip of the force of gravity pushes the temperature of material to tens of millions of degrees, triggering nuclear fusion between the hydrogen and its isotopes. The reaction produces a titanic outpouring of heat that propagates everywhere in the form of an irresistible flow of photons and neutrinos.

A blinding flash of visible light flares up in the deepest darkness. The universe is still shrouded in darkness, but the new light has just begun to cover the seemingly endless distances, and soon it will be joined by a myriad of other light sources that will be kindled everywhere.

When we get to the fifth day and 200 million years have passed, the first star is born.

And So We Go Out to See the Stars

There is no verse more powerful than that chosen by Dante to close the *cantica* of the *Inferno*. The hendecasyllable is a distillation of that feeling of consolation which the star-studded sky has inspired in humanity since the beginning of time. The same mood that will inspire an equally striking opening of Giacomo Leopardi's:

O faint stars of the Great Bear, I never thought
I'd return to contemplate you again
Sparkling above this paternal garden.

Having experienced all manner of fears and risks in the shadowy world of the Inferno, in the darkness that hides anguish and tortured flesh, or at the culmination of a bitter reflection on a life lived differently from the one he had imagined, seeing the stars again fixed in the firmament is reassuring and calms his fears. With its apparent constancy and even immutability, the starry sky protects us from the fear of change and catastrophe: it comforts us and caters to our childish desire for stability.

And yet if we observe them closely or investigate the mechanisms that agitate the innermost structure of these marvellous stars, we encounter material processes of such tremendous violence that it is hard to find more turbulent and unstable systems than these.

A star such as our Sun seems gigantic to us, with a radius

a hundred times bigger than that of the Earth, which by comparison seems to be an insignificant dot. And yet the Sun is a yellow dwarf, a medium-sized star, one of the many that abound in our galaxy. It has nothing to do with the gigantic counterparts in this category such as the major star of the Eta Carinae system, a monster that has a mass almost a hundred times that of the Sun. But as we shall see, in the world of stars, being reduced in size has significant evolutionary advantages.

The Sun is an almost perfect sphere of incandescent plasma, composed largely of hydrogen and helium: it is endowed with a magnetic field, and rotates on its axis every twenty-five days. The surface temperature is close to 6,000 degrees, but it rises to more than 1 million degrees within its interior. The origin of this vast amount of energy resides in the mechanisms that ferment at the heart of that great ball of ionised gas. This immense concentration of matter produces a gigantic gravitational force, which compresses the plasma strata; the temperature rises steadily as we approach its innermost shell; at the heart of the star the temperature exceeds 15 million degrees, and in this environment thermonuclear fusion reactions are triggered.

The process of fusing together two light nuclei yields an enormous quantity of energy. The final bound state is lighter than the two nuclei were before and the difference in mass is transformed into the energy that is developed by the reaction.

There is a problem, however, inasmuch as it is anything but simple to fuse together, for instance, two protons or hydrogen nuclei. They both have a positive charge and violently repel each other when one tries to bring them into contact with each other, that is to say, to that distance at which the gravitational attraction would manage to overcome the repulsive effect of electromagnetism. This can only be achieved by exploiting the collisions that arise from extreme temperatures and pressures.

Within the interior of the Sun, beneath the pressure of the innate force of gravity, these conditions are realised – or to be more precise they get close enough to succeed in triggering the phenomenon. The majority of the protons do not take part in the fusion – only an infinitesimal fraction which, due to the effect of quantum fluctuations, succeeds in overcoming the potential barrier. The phenomenon involves a mass of hydrogen large enough to allow the production of a tremendous amount of energy, but small enough to allow the star to shine for billions of years.

In the heart of the Sun, hydrogen nuclei and their isotopes, deuterium and tritium, fuse together to form helium nuclei. The energy liberated by the reaction appears in the form of high-energy neutrinos and photons. The former have no problem crossing the immense, incandescent sphere and fly freely to colonise the most remote corners of the universe. The photons might dream of following suit, but remain held in seemingly endless captivity. Crossing the hyperdense matter that surrounds them, the photons collide with and are continually absorbed and re-emitted by the material they encounter on their route. In this way their energy is reduced and their initial direction lost. They will wander in this infernal labyrinth for millions of years, because the cycle will be repeated countless times before they can succeed in escaping from this vice-like grip. Finally, one day when they already think of it as a hopeless task, they will emerge almost by chance from the surface and will at last be free. From now on they will be able to cover endless distances: they will fly to distant points at the speed of light, to heat up and shed light on everything around them.

The thermonuclear reaction holds everything, the entire system, in precarious equilibrium. In the depths of the Sun an unequal struggle takes place between gravity and the strong force. The weakest of the interactions, the effects of which have long been ignored, takes its revenge and forces into collision the first in the class, that strong interaction that used to look down on it. After it has pulled all the hydrogen that was wandering in the vicinity into its influence, and has herded it and organised it into the perfect spherical geometry of the Sun, it knows that it has become invincible.

A terrible pressure crushes matter and seeks to shatter it into its elementary components. The protons confined and constrained by the fusion momentarily manage to escape from their destiny; the huge quantity of heat that is unleashed with the formation of helium nuclei tends to make the plasma expand and resists the grip of gravity. An equilibrium is created, though one that is intrinsically unstable because sooner or later the hydrogen will be exhausted. But this battle could last for billions of years.

This most turbulent of environments, devastated by convection currents, immense vortices and gigantic projectile plasma, when seen from some distance will appear to us to be a beneficent and reassuring star, and its praises will be sung by everyone on Earth as the mainstay of the order that prevails in the world.

For millennia we will remain ignorant of the furious struggle taking place within it. It's a battle on an epic scale, yet with an outcome that is taken for granted, because the name of the victor is already well known – and it is equally well known that the collapse of the defeated, when it comes, will be catastrophic.

The clash between Zeus and the gods of Olympus against the Titans led by Cronos lasted for ten years. With the help of his thunderbolts, the new arms forged by the Cyclops, and the slingshots of their allies the Hecatoncheires, the hundred-handed giants, Zeus defeated the Titans and plunged them into the deep darkness of Tartarus. The struggle to the death between gravity and the strong nuclear force that occupies the centre of the battlefield in the core of the Sun will last much longer. It will take ten billion years for the available hydrogen to be consumed, but when this finally happens nothing will be able to oppose gravity, and disaster will follow.

The Heroic Age of the Megastars

The first stars that shone in the universe, 200 million years after the Big Bang, were astral bodies of a very particular kind. It's thought that they were truly gargantuan, between one and two hundred times bigger than the Sun, which is why we call them *megastars*. They were formed in the profound darkness of the era of obscurity, taking tens of millions of years to aggregate the immense quantity of hydrogen required. The hunt is on to find one still shining in a remote corner of the universe, without any success to date.

After recombination, the ordinary matter of the universe is made up of atoms (hence it is completely neutral) and is still undergoing a process of cooling. Gravity slowly concentrates it around the knots of greater density in the distribution of dark matter that envelops the enormous cloud of gas. Irregularities translate into zones of intense gravitational attraction, which thus form agglomerations of matter on an ever more imposing scale.

The first megastars did not originate in isolation, but aggregated into clusters of various sizes, organising themselves into extended families. This locally asymmetrical spatial distribution will be reflected in the subsequent formation of galaxies.

They are extremely unlike current stars, not just because of their size but because they are composed exclusively of hydrogen and helium. These megastars are completely lacking in heavier elements, for the obvious reason that these have yet to be formed. The synthesis of nuclei of carbon, nitrogen and oxygen that will provide the indispensable components for the emergence and evolution of more complex structures such as galaxies and planets will only occur in the innermost strata of these new stars.

In dwarf stars such as the Sun, the descendants of a long chain of generations of primeval stars, these elements are present but do not participate to any significant degree in the nuclear processes dominated by the proton–proton chain. And on the other hand, stars more massive than the Sun, reaching far greater internal pressures and temperatures, can trigger other nuclear fusion reactions involving heavier elements. In particular, at sufficiently high temperatures the nuclei of carbon, nitrogen and oxygen can act as catalysts for the fusion of hydrogen and increase its efficiency. This same process constitutes a limit to the dimensions of the more massive stars in the current universe. For a mass of around one hundred and fifty times that of the Sun, the nuclear reactions linked to the carbon–nitrogen–oxygen chain would have such speed as to rapidly bring about the destruction of stellar structure.

This limit does not apply to megastars: the speed of the single proton–proton reaction makes it possible for true monsters to be built, reaching a size of up to three hundred solar masses. The bigger the star, the quicker its consumption of combustible fuel. For stars the notion that 'small is beautiful' applies, inasmuch as there are considerable advantages to having more limited dimensions. The Sun may burn slowly for billions of years, whereas the supergiants that look down on it on account of its size will have a relatively brief life of no more than a million.

The superstars that started to shine in the early universe 200 million years after the Big Bang are impressive entities, with a dominant, extremely luminous but short-lived presence. They bring the era of obscurity to an end with their light, but their existence is as ephemeral as that of fireflies in spring.

Megastars follow a fixed succession of events, from generation to generation, and when they come to the end of their lives they explode, scattering all around themselves the new forms of matter they have forged in their immense nuclear crucibles. In this way the universe is enriched with elements such as carbon, oxygen and nitrogen, and gradually accumulates other, increasingly heavy ones that will also modify the nuclear reactions of subsequent stars. The stars that will utilise the material distributed throughout space by the megastars will be smaller and less bright than their gigantic ancestors, but they will survive for longer and give rise to more complex transformations which require, above all, a much greater length of time.

Like the supersized pachyderms of the Jurassic period that brought in their wake smaller and more agile animals, the megastars that would be extinct in a couple of hundred million years gave way to new generations of stars that were smaller but better adapted to survive.

To gather signals from this dark and silent epoch in which the first stars were formed is one of the challenges facing modern radioastronomy. The only radiation emitted by the large clouds of gas being aggregated into superstars is what is known as the '21-centimetre line' of neutral hydrogen. This is a characteristic electromagnetic signal emitted by hydrogen in the microwave region, and its discovery would provide unequivocal confirmation of having succeeded in penetrating into the opacity of the cosmic dark age. It consists of a very weak signal born of a prohibited transition of the hydrogen atom, a very rare phenomenon that may be observed only when enormous quantities of gas are being investigated. Radioastronomers have managed to reconstruct it through soundings of the great nebulae of hydrogen present in our galaxy, but all attempts at identifying it in the background noise of the universe have failed.

If it is discovered, a similar map to that of cosmic background radiation could be charted, furnishing us with a very precise picture of the scattering of matter in the era of obscurity: we would be able to see in detail the mechanism behind the formation of superstars, and would be better able to understand the role that the phase of reionisation had in the formation of the galaxies.

With the frenetic life and death cycle of the great primeval stars, a new phenomenon emerges: the light emitted by them is so intense that when it floods over the hydrogen distributed in the surrounding space, it ionises the atoms of the gas, tearing away from them the clinging electrons. The phenomenon is even more violent at the death of a megastar, when a blinding glare signals the end of the nuclear fuel. Very slowly, the greater part of the material present in the universe becomes completely ionised, returning to the state it had abandoned during the recombination era, 380,000 years after the Big Bang – and we see a progressive increase in opacity. It is the age of reionisation which begins a few hundred million years after the appearance of the first megastars.

For a long period the universe turns dark, in a continuous alternation of light and darkness that seems to be endless. Now the universe is full of enormous, very luminous stars, but it is no longer transparent. The free electrons interact with the photons emitted by stars, attenuating and capturing them, thus preventing them from transmitting light across vast distances. The universe thus plunges back once again into total darkness.

The process will continue for a few hundred million years, the time required to ionise all of the hydrogen gas. Matter has now gone back to being plasma, to a state similar to the one that provoked the dark age and that could, in theory, absorb all of the light being produced. But the universe continues to expand, and the density is progressively reduced until it reaches such a low point that, the process of ionisation having come to an end, everything becomes transparent again. From this point onwards the whole universe is pervaded by a hot, ionised gas, but it is so rarefied that light passes straight through it.

Finally, before the universe has reached its billionth year, light has prevailed over darkness. It was a hard-fought struggle, and there were moments when it seemed as if light would be extinguished forever. But now it is triumphant – and this time its success will be lasting.

An Incredible Cosmic Firework

The nuclear processes unleashed inside the megastars lead to the formation of increasingly heavy elements. Carbon, nitrogen, oxygen, and all the rest of the elements, including iron, accumulated slowly in the innermost strata, imprisoned by gravity. At the end of their life cycle, the structure of these massive stars was shattered by immense explosions that sent everything flying into surrounding space. After numerous such destructive cycles, from out of this stellar dust that was rich in heavy elements including many kinds of metal, other stars and other planets emerged, such as our Sun and Earth.

The violent phase in which stars die, producing truly spectacular effects, has a decisive role to play in the formation of our solar system, and it is worth describing it in more detail at this point.

How stars end depends largely upon their mass. Heavy stars with a mass ten times that of the Sun produce in their core temperatures and density levels that are monstrously high. At the heart of these giants the heat exceeds billions of degrees, and at these inconceivable temperatures the fusion reactions involve all of the elements. With the passage of time the lighter components – hydrogen and helium – are consumed, and the heavier elements produced by more complex reactions – carbon, nitrogen, oxygen, and so on – begin to melt. When we begin to see silicon melt and iron produced, the process stops. There are no further reactions possible, and the heart of the star that is producing no more energy collapses catastrophically.

Beneath the unrelenting pressure of gravity, the central nucleus suddenly contracts to become hundreds of thousands of times smaller – and the star explodes. All of the layers surrounding it find themselves suspended in the void, and the fierce power of gravity plunges them back towards the nucleus that has become an extremely small, tremendously compact object. The overwhelming impact upon the nucleus, and the nuclear reactions that ensue, scatter all of the material outwards in every direction. An enormous mass of gas, equivalent to that of many Suns, produces an immense shock wave that shoots through space at more than 10,000 kilometres per second, remaining visible for hundreds of years. The clouds of gas, rich in heavy elements and chemically diverse, will cover vast distances and supply the foundational material for new aggregations.

Just as the power of Zeus flings the Titans into the abyss, so the force of gravity, infuriated by all the time lost spent opposing nuclear force – the force that has so far prevented its triumph – takes its revenge and celebrates its victory with a horrifying silent scream that tears the star to pieces and flings it in fragments into space at incredible speeds.

A dazzling flash of light crosses the heavens. It is so spectacular that when ignorant earthlings will in due time catch sight of it thousands of light years away, they will think that this dot of light that appeared suddenly in the sky is not merely a sign of the death of a star but instead indicates the birth of a new star which they will call a *supernova*. The phenomenon will be the source of universal astonishment, registered throughout history as either a dismal augury or an auspicious flicker, depending on the needs and circumstances of the observers.

All of the nuclei that make up our bodies – the calcium in our bones, the oxygen in water, the iron in haemoglobin – have traversed this tempestuous and terrible past. Now the formative atoms submit docilely to chemical and biological reactions which guarantee our existence. If only they could tell some stories about their eventful childhood ... or speak about the nightmare of such a traumatic birth: first produced in the extreme heat and pressure at the heart of a star, then flung at incredible speeds into the absolute void, for billions of years, waiting for the coming of a new aggregation.

Explosions of supernovas are among the most disastrous occurrences in the universe and provide us with a precious source of information on the dynamics of stars and the construction of galaxies. The phenomenon leaves an immense amount of energy in its wake, adopting various forms. Most of this energy is emitted in the form of neutrinos: a grotesquely large flow of these extremely lightweight particles illuminates the entire universe every time a supernova explodes. Fortunately, neutrinos are gentle and delicate, and the only trace of their earthly passage that they leave behind are some innocuous flickers in the enormous detectors dedicated to their pursuit. A significant amount of the energy is deployed in the acceleration of the shock wave that pushes material all around it. The rest consists of gravitational waves and electromagnetic radiation of all frequencies: the light that produces the visible glow that even we can see, but above all high-energy photons, X-rays and gamma rays which together with the charged particles are accelerated by the shock wave and hurled across immense distances. These are phenomena that may last for weeks or even for months. Others, linked to radioactive isotope decay produced by the gas, may last for decades.

The explosion of a supernova is one of the most awesome natural occurrences that the human mind has been able to witness and comprehend, but it is just as well that we are not observing it close-up. The effects of the radiation could be lethal for many, if not for all, of the species on our planet. Luckily for us, the massive stars that we can predict will quit the scene with such spectacularly pyrotechnical finales are quite rare – and all safely distant from us.

The closest to us is Betelgeuse, a reddish star that's visible to the naked eye, just above Orion's Belt. It is a red giant of enormous diameter, weighing ten times more than the Sun. It is a star of such incredible dimensions that if we positioned it where the Sun is, it would fill the solar system, reaching almost to the orbit of Jupiter. This star is approaching the end of its life, but we cannot predict exactly when this will be. It could explode any moment or continue to shine for thousands of years, but we can be sure that when its time comes it will go out with quite a show. The after-effects of its explosion will illuminate nights on Earth for months, like a constant full moon. The tremendous fireworks it will cause should not present any dwellers on our planet who might have survived until then with any imminent danger: fortunately the star is 600 light years away, a distance that should allow them to enjoy the show in complete safety.

So what about our Sun: how will *it* come to an end? It

is much too small to explode catastrophically, and when its time is up it will bow out discreetly. This would still be a worrying enough event, if it wasn't for the fact that it is not due to happen for quite a while yet. We should be fine, given that it will take 5 or 6 billion years for the Sun to exhaust its supply of hydrogen. When its end does come, the reactions will begin that involve the heavier elements, and at this critical point its innermost core will heat up and the Sun will increase unstoppably in size, becoming a red giant. Its dimensions will increase rapidly until they reach and vaporise in turn Mercury, Venus and Earth. This particular development should not concern us, since long before it happens the Sun will have grown by 40%, meaning that the glaciers at the poles will have vanished and the oceans become desert landscapes. Any form of life on Earth will long since have been rendered impossible.

Having reached its end, the Sun will expel its outer strata of gas and be transformed into a planetary nebula. Slowly its innermost core will free itself of its crust, and an object similar in size to the Earth will emerge. Extremely dense, fiercely hot and luminous, the Sun will have become a *white dwarf*, which is to say a small brilliant body made up of carbon and oxygen nuclei that are completely ionised and protected by a compact shield of electrons strong enough to impede ultimate gravitational collapse. This little astral body will continue to cool, perhaps for billions of years, until it becomes a *black dwarf*, which is to say an inert body, completely invisible, devoid of any trace of its former splendour.

The Fascination of Black Stars

Stars much bigger than the Sun, when they have exhausted their combustible nuclear fuel, are transformed into even more exotic entities: if their mass is equal to between ten and twenty solar masses, they form extremely dense *neutron stars*. These small spheres with a radius of between 10 and 20 kilometres contain one and a half times the mass of the Sun.

Neutron stars are formed when the gravitational collapse is so violent that all of the nuclei of the elements of which a star is made are smashed into a pulp of protons and neutrons. The electron gas that in white dwarfs acts as a protective shield is shattered in an instant. The force of gravity in such massive entities is so great that the electrons and nuclear matter is so compressed as to trigger capture reactions on the part of the protons, which all become transformed into neutrons. An extremely compact, monstrously dense body emerges, similar to a gigantic atomic nucleus, made entirely of neutrons densely packed together by the strong force. It is matter that is so dense that the mass of a mountain the size of Everest with the same density could be contained in a teaspoon.

If this wasn't already impressive enough, the small sphere rotates on its axis at frenzied speed. Neutron stars have been identified that need just a few thousandths of a second to make a complete rotation. The surface strata of these stars which rotate at hundreds of revolutions per second, move at speeds that can easily exceed 50,000 kilometres per second.

This phenomenon arises from the extreme contraction produced during its collapse. The mother star's movement, its slow and gentle rotation around its own axis, is accelerated by the conservation of the angular momentum. If the original rotation could be measured in terms of weeks or months, when the radius contracts from millions of kilometres to just a few dozen, the frequency rises to a hundred revolutions per second. It is like when an ice skater suddenly folds her arms tight, and the pirouette she is performing becomes much faster and more spectacular.

The rapid contraction of its dimensions, linked to the gravitational collapse, also amplifies to an enormous degree the original magnetic field. Those gigantic lines of force that surround the huge star are now forced to fold into a small, compact nucleus where their density causes an explosion. Neutron stars produce extreme magnetic fields, billions of times stronger than those of ordinary stars.

When the magnetic axis of neutron stars does not correspond perfectly with the axis of rotation, electrons and positrons that have remained free on the surface of the star become subject to acceleration towards the poles and produce a powerful beam of electromagnetic radiation that rotates with the same frequency as the star. If the Earth finds itself within the emission cone of this very special kind of radio station, we will be able to record a radio signal with an extremely regular pulse, a super-precision timepiece, a sort of incredibly powerful lighthouse which emits radio waves instead of light. We will have discovered a *pulsar*.

The Singularity of Black Holes

When the mass of a star is truly abnormal, upwards of thirty times that of the Sun, its collapse results in the formation of a *black hole*. Not even neutrons can withstand the degree of gravitational force exerted, and end up in pieces; even their elementary components become so extremely compressed that their residual mass is concentrated into a virtually infinitesimal volume.

Following this pattern, systems emerge at the core of which there are laws of physics that remain inscrutable to us, and that allow for the cramming of five to fifty solar masses into a small inaccessible space with a diameter of just tens of kilometres.

It may in this way remind us of one of the most commonplace nightmares – of plunging irresistibly into a bottomless well. Or perhaps because of inherited fears from our remote past of being savaged and devoured by wild beasts, the very idea of black holes seems to immediately trigger a kind of echo of an ancestral panic response.

Until a few years ago black holes interested a few thousand specialists at most, who in their discussions at academic conferences were surely oblivious of the coming explosion of popular interest in such an arcane subject.

That our heavens might contain *dark stars* is an idea that has been around for at least several centuries. The first to propose their existence, in 1783, was the natural philosopher and eminent scientist of his time, John Michell. Extrapolating from the corpuscular theory of light developed by Newton, Michell was able to readily imagine stars that were so compact and massive as to produce gargantuan gravitational attraction, so exceptionally powerful as to imprison forever the light that was emitted by their surfaces. The light particles would behave like stones thrown from the surface of the Earth, describing parabolic trajectories that would inevitably return them to the ground.

Michell's theory was so far ahead of its time that for almost two hundred years no one took it seriously. A first breakthrough moment came in 1916, soon after Albert Einstein had recently published his theory of general relativity and Karl Schwarzschild, a German physicist who had volunteered during the First World War and seen action as the commander of an artillery battery on the Russian front, had managed to have sent to him the article that was about to make history. In a short period of time, using a different system of coordinates, Schwarzschild succeeded in finding an exact solution to equations for which Einstein himself had only found approximate ones.

With this new approach space-time assumed a spherical symmetry, and for every mass it was possible to define a radius, known as the Schwarzschild radius, below which a singularity would occur: a bending of space-time to such a degree that the very photons would not be able to escape. The result was so curious that neither Einstein nor Schwarzschild himself dared to write it up, or even to imagine that behind the mathematical formulas a whole new class of celestial bodies might be hidden.

We would have to wait until the 1960s before the term *black hole* was coined. It was introduced in 1967, with a certain degree of irony, by the American physicist John Wheeler, who was among the first to intuit that we were dealing with real astronomical objects, opening up a whole new field of research. From this moment onwards, modern astrophysics has been deeply marked by the study of black holes and the hunt for all possible clues that might indicate their presence.

The 1970s brought the foundational theoretical contributions of Roger Penrose and Stephen Hawking, and the first indirect observations of candidates for black holes. The catalogue of the latter has grown year on year, until the surprising discovery of the presence of supermassive black holes at the core of most elliptical or spiral galaxies. Finally, the reader might remember, it was due to a collision between black holes, consisting of around thirty solar masses, that in 2015 the first gravitational waves were recorded by the huge interferometers of LIGO in the United States.

Black holes can be 'observed' indirectly, with the aid of signals that arise from their interaction with forms of ordinary matter. When they orbit in the neighbourhood of a massive star, their powerful tidal forces strip their unfortunate neighbour of substantial amounts of matter: the ionised gas, accelerated by the gravitational field of the black hole that is about to swallow it, forms disc-shaped growths that emit radiation of many different wavelengths. What makes the pyrotechnic display even more spectacular is the fact that powerful jets of matter, emitted from the polar regions, are frequently registered travelling through space at close to the speed of light.

Black holes are therefore a new kind of celestial body – one that is quite rare but which is nevertheless present in many areas of the universe. Today we know that they are very different from each other, not just in dimensions and properties – stationary or rotational, neutral or charged – but also because of the dynamics behind their formation and evolution.

They may form as the result of the gravitational collapse of supermassive stars. But they can also come about when neutron stars collide with each other, or reach critical mass by absorbing matter from ordinary stars with which they interact in binary systems.

A Fusion Worth More than Its Weight in Gold

As well as giving rise to black holes, the collision of neutron stars can produce other astonishing effects.

Imagine an enormous cloud of gold and platinum, with a mass hundreds of times greater than that of the Earth. This was the incredible spectacle seen by astronomers when they recently concentrated their instruments on a zone of the heavens close to the Lyra constellation. A veritable 'cosmic heavy metals factory', it was the result of a single catastrophic event: the collision between two neutron stars.

It is August 2017 and for the first time, after a few days, the two American interferometers of LIGO and the Italo-French one of VIRGO, near Pisa, are operating together. They are searching for gravitational waves produced from the fusion of black holes, and they immediately record an event similar to that of the first discovery in 2015. Then, after just three days, they detect a new signal, an anomaly, strangely out of the ordinary – less intense but much longer-lasting. It is the characteristic signature of gravitational waves produced by the fusion of neutron stars.

These are not the ultramassive bodies that had been the origin of the first signals; two neutron stars, when they meet, end up by merging together in a catastrophic crash, and when they spiral around each other and reach speeds close to that of light, they warp space-time and produce a signal of gravitational waves lasting tens of seconds.

All of this has happened at a distance which in cosmic

terms is rather modest: just 130 million light years away, compared to the 1.4 billion of the first sensational discovery.

The fact that this time VIRGO was also fully operational made triangulation possible. With three instruments active it was possible this time to identify the source, and the alert sent to seventy observatories spread across five continents, as well as in space, produced a mass of data. The gravitational wave signal was accompanied by high-energy photons and sequences of lower-energy electromagnetic emissions that would last for weeks.

It was immediately understood that the flash of gamma rays detected a few seconds later by other instruments, such as FERMI – a special telescope in orbit around the Earth – came from exactly the same region and was connected to the same phenomenon. In all probability this was the signal that a black hole had been formed from the collision.

The momentous events of the 17 August 2017 constituted the first spectacular breakthrough of *multimessenger* astronomy. The same phenomenon is being studied using the signals emitted in the full range of wavelengths of the electromagnetic spectrum and by gravitational waves, and from this it is possible to create a much more detailed understanding of such occurrences.

We now know that when two neutron stars merge together they produce gravitational waves; and we have understood where the gamma-ray bursts come from, the origins of which we'd had so many questions about. In the end we were rewarded by an incredible surprise that emerged in the weeks after the identification of the first signal: astronomers identified, in the residues left behind by the fusion, a small nebula of heavy materials. An enormous quantity of precious-metal dust, a gigantic mass of gold and platinum produced by the collision and expelled at extreme speeds into surrounding space spectacularly confirmed the theory that elements heavier than iron could only be formed by catastrophic events of this kind.

Once again, we have the experience of discovering phenomena of such disproportionate extreme violence hidden behind the appearance of cosmic equilibrium that at first sight is so calm and reassuring.

With the description of these extraordinary events, our story has reached the end of the fifth day. The universe is populated by a myriad of stars that in a succession of generations have spread immense quantities of gas and the dust of heavy elements throughout the universe – and among them lurk neutron stars and black holes. Five hundred million years have passed since the birth of the universe, and the first galaxies are forming.

Day Six: And Chaos Disguised Itself as Order

At the beginning of the sixth day the universe shines with a myriad of gigantic stars. They reproduce, from generation to generation, in temporal cycles that compared to cosmic ones are swiftly concluded. Every time one of these giants dies, the great cloud of ionised hydrogen and helium that surrounds them is enriched by increasingly heavy elements, contributing to great nebulae of gas and dust that in turn will give rise to new generations of smaller, longer-lasting stars.

Gravity acts slowly upon these agglomerations of matter that have formed around the great clumps of dark matter; by far the most predominant, in terms of mass, these agglomerations generate veritable potential wells towards which stars, gas and dust launch themselves. Everything seems to run headlong towards this nothingness, an invisible heart of darkness that unstoppably attracts everything. As a result of the shocks that are created in this compression, the gas becomes hotter and increases the pressure that will attempt to oppose ultimate collapse. Most of it is concentrated around the centre of the halo of dark matter, where density increases and everything else gravitates around it.

The conservation of angular momentum prevents the stars and agglomerations of matter from being drawn, directly, into the central hole; the underlying symmetry compels them to turn, slowly, around the central nucleus and a rotation disc is formed, a vortex similar to that of hurricanes. This is how a galaxy is born.

We are falling, unstoppably, there can be no doubt, and the fall is unavoidable. We are being swallowed up by a terrible whirlpool; our worst nightmare is coming true. Our fate has been sealed, the chaotic and powerfully dynamic mechanism that is governing events leaves us no hope. The duration of the catastrophe is extremely prolonged, not only in relation to our individual lives but also compared to the life of our species, which has inhabited the Earth for just a few million years. The life of a galaxy develops on a scale of many billions of years; there will be plenty of time in which to construct solar systems and planets, as well as forms of life capable of asking how all of this works.

Chaos has effectively disguised itself as order, wearing a mask of balance and harmony – and this great subterfuge keeps us calm, reassuring us for millennia.

Spira Mirabilis

The name of our own galaxy, the Milky Way, includes a literal translation of the Greek word *galaxias* ('milky') from which we get our generic term for a galaxy. The name encodes a myth of origins, connected to one of the many indiscretions of Zeus. Having been smitten by Alcmena, the king of the gods assumes the appearance of her husband, makes love to the beautiful mortal and in doing so impregnates her. From this coupling Heracles will be born – and immediately abducted to Olympus by Zeus. There he places him in the lap of his sleeping consort, Hera, so that the infant will taste the milk of the goddess, and with that become immortal himself.

But the little creature, a kind of changeling in reverse, even as a baby incapable of keeping in check the physical exuberance that will one day lead him to perform legendary feats, attaches himself to her breast with excessive force and suckles voraciously. Abruptly awakened, Hera pushes off this unknown infant, and the milk that spurts from the nipple of the goddess fills the sky with whitish droplets which are immediately transformed into tiny stars, while the ones that fall to Earth germinate as lilies.

Our Milky Way is an agglomeration of stars, dust and gas held together by an enormous halo of dark matter. It is a great spiral galaxy, a gigantic cosmic Catherine wheel, organised into brighter arms, in which newly formed stars are concentrated. It contains more than 200 billion stars and everything revolves around its dense central region. At its core, the concentration of matter is so high that a kind of bar of constant density develops, from which the type *barred* spiral galaxy gets its name.

Its shape conforms to the geometry of spiral growth, a curve that is found in many natural processes. Starting from the centre, the radius grows regularly with the angle, forming the beautiful geometry typical of certain shells such as that of the nautilus. Descartes was the first to describe its function, and Jacob Bernoulli was so taken with it that he dubbed it *spira mirabilis*, the marvellous spiral, and specified that one should be carved on his gravestone.

Unlike what happens in the solar system, where the velocity of the planets decreases with distance from the Sun, here everything orbits around the galactic nucleus with almost identical speed – about 200 kilometres per second, which is to say, an astonishing 700,000 kilometres per hour. We have already seen that this almost constant orbital speed is one of the more obvious indicators of the hulking presence of dark matter. In fact, what we call our Milky Way is only a small part of our galaxy.

Dust, gas and stars, that is to say visible matter, are distributed on a flat disc, about 100,000 light years in diameter and 2,000 light years thick. Our Sun, dragging its dependent planets in its wake, orbits at a distance of around 26,000 light years from the centre of the galaxy, and despite its considerable speed must labour for more than 200 million years in order to make a complete orbit. Everything is immersed in an immense spherical halo of dark matter that is estimated to have a diameter of a million light years. The luminous part is almost insignificant compared to the enormous cloud of invisible and mysterious matter that seeps everywhere and surrounds everything, contributing around 90% of the global mass.

Galaxies, Clusters and Collisions

The phase in which the great galaxies are formed covers a period of considerable length in the life of the universe. In fact, the first aggregations start to form 500 million years after the Big Bang, and continue for up to 3 to 4 billion years, while smaller, more compact galaxies will continue to form for further billions of years.

The Milky Way has dimensions that are far larger than average. Given the volume that it occupies and the number of stars that it contains, we are in fact justified in thinking of it as a giant galaxy. That said, there are other, truly enormous galaxies that make the Milky Way look ridiculously small. One of these is IC1101, a supergiant galaxy containing over a 100 trillion stars and boasting a diameter of 6 million light years.

The total number of galaxies in the universe has been calculated by extrapolating from those observed in a small portion of the heavens which had seemed to contain none. The result is pretty impressive: the most recent estimates suggest that there may be some 200 billion galaxies in existence. And this excludes those that are too small or not bright enough to be observed across such huge distances.

Together with spiral galaxies, the most common are elliptical: in these the stars are distributed in an ovoid volume which is almost spherical. These two types account for about 90% of the total number of galaxies, with the other 10% consisting of irregular shapes.

The most extravagantly irregular of these are frequently the smaller galaxies. Among these are ring structures in various configurations – not to mention the even more peculiar ones that have been compared to the outline of a penguin or to the letters of the alphabet. Frequently the quirkier shapes are the result of collisions between galaxies. During the impact it is highly improbable that a single star will collide with another celestial body, but the strong gravitational interactions that result from the close encounter destroy the ordered structure of the system, which then adopts the most bizarre shapes. It is believed that all galaxies first emerged in disc form, and that elliptical ones are the result of merging or even the cannibalising of satellite galaxies.

Around the Milky Way we find two giant galaxies: the nearest is Andromeda, while the Triangulum galaxy is only a little further away. The three galaxies together belong to the Local Group of galaxies, around which satellites such as the Large and Small Magellanic Clouds gravitate. There are more than sixty of these satellite galaxies, many of them dwarf ellipticals, some of them truly minute, containing just a few thousand stars.

Our Milky Way and the Andromeda galaxy seem to be set on a collision course. The distance between them is considerable – 2.5 million light years – but we should not underestimate the speed at which they seem to be converging: 400,000 kilometres per hour. Basically, there is a possibility that in 5 to 6 billion years the two great galaxies will be involved in a cosmic crash of an unimaginably spectacular kind. As they approach each other they will become subject to a prolonged period of turbulence, during which tidal forces will distort in an irreversible way the two marvellous spirals, perhaps producing in the process a single huge structure. For a while the Triangulum galaxy would remain a spectator to these events, before being transformed into a satellite of the new galaxy that would emerge from the fusion of the two giants, with the possibility that it too will eventually merge with the new, enormous aggregation.

Local groups can be made up of dozens of galaxies; if we get beyond a hundred the terminology changes: we no longer talk about groups but about clusters. Groups, clusters and isolated galaxies in turn form even more gigantic structures called superclusters. This hierarchical organisation is quite common and can be found everywhere. The Local Group of the Milky Way, for instance, forms part of the Virgo supercluster, a vast system which includes almost 50,000 galaxies. The different superclusters are connected with each other by strands of galaxies across empty areas of great extent. This hierarchical type of organisation ends up by forming a sponge-like superstructure, with enormous empty bubbles interspersed with areas with a high density of galaxies. This is the large-scale structure of the universe.

The Dark Heart of Our Milky Way

Looking southwards on a clear summer night, the heart of our galaxy can be seen with the naked eye, just above the horizon in the constellation of Sagittarius. We will not be able to make out many stars, but if the air is clear and we are far enough from sources of light pollution, we will see a kind of diffuse glow. It is what remains of the light from a large concentration of stars, attenuated by dust that thickens around the galactic centre. To get a clearer picture we need to use telescopes capable of penetrating through the dust, such as infrared ones, or those that use a kind of X-ray radiography.

Observations using such instruments have made it possible to highlight the enormous concentration of stars in the centre, and have led to a disturbing discovery. When the orbital rotation speed of some of these stars was measured, it became immediately apparent that something was wrong, as they all seemed to be moving at speeds significantly higher than expected. When it was decided to measure, for months, the movement of dozens of these stars very close to the centre of the galaxy, astonishing speeds were recorded. One of these stars was found to be rotating at 5,000 kilometres per second.

When dozens of stars were seen orbiting around nothing, at speeds implying a tremendous gravitational attraction, there was only one possible conclusion: at the centre of our galaxy there is a huge mass concentrated into an invisible and gigantic object weighing 4 million times more than the Sun. We have just described Sagittarius-A*. At its deepest, darkest core, our apparently placid Milky Way harbours this monster. And it is at this point that our worst ancestral nightmares are realised: we are plummeting into a bottomless gravitational well that sooner or later will swallow everything.

Sagittarius-A* is a black hole of enormous mass, with a Schwarzschild radius of around 12 million kilometres. Its density is certainly high, but not even remotely comparable to that of black holes of stellar origin, which are much less heavy but also tiny in size. It belongs to a new class: supermassive black holes. It has properties which differ radically from those of its fellows, which are the last stage in the evolution of big stars. Compared to Sagittarius-A*, the black holes with thirty or so solar masses that produced the first signals of gravitational waves seem like diminutive and almost tame objects.

As chance would have it, the black hole that is closest to us is located precisely there, at the centre of the constellation in which Greek myth placed Chiron – half-man, half-horse, and the most skilful of archers. Chiron was the freakish offspring born of the unnatural coupling of Cronos, in the shape of a stallion, with the nymph Philyra. Abandoned by a mother disgusted by his appearance, and educated in all the arts by Apollo, he became the most civilised of the violent and bestial centaurs. He is the prototype of Sagittarius, a symbol of man who through knowledge and culture overcame his animal nature: Chiron, the great practitioner of the medical arts, that legend has decreed was a master of learning and a mentor of heroes, starting with Achilles.

Sagittarius-A*, like Chiron, can help us understand a world that is hostile and apparently full of danger. Studying the behaviour of supermassive black holes, those turbulent regions in which matter interacts in extreme conditions, may hold the key to understanding very important matters whose nature still escapes us. It is for this reason that so many telescopes and instruments of every kind are focused on it, accumulating increasingly astonishing data.

We have discovered that gas and dust, plummeting towards the black hole, is heated to millions of degrees and emits radio waves as well as infrared radiation. Sagittarius-A* probably has a magnetic field, and traces of an accretion disc have been detected, that is to say a kind of ring formed of matter that has been torn from the nearest stars and now orbits around it. Signals have been picked up that seem to indicate relativistic jets at the poles: a sort of hiccup or regurgitation occurring when the monster has ingested large quantities of dust and gas, and expels a part of it, thrusting it towards the poles so violently that it moves at almost the speed of light.

Finally, in the last of a series of surprises, when observing a cluster of seven stars three light years away, astronomers have discovered another black hole. The cluster is held together by this object, which is as heavy as 1,300 suns, and it all orbits around Sagittarius-A*. It is the first black hole with an intermediate mass discovered within the inner region of our galaxy, and its presence may give an indication of the abnormal growth of Sagittarius-A*, surely due in part to the cannibalisation of other black holes with comparably large dimensions. The most recent discovery of another dozen or so black holes surrounding it has served to strengthen this hypothesis.

Being so close to us, the central core of the Milky Way provides an ideal laboratory to stress-test general relativity and to study the phenomena that occur in areas prone to high degrees of spatio-temporal distortion. This is why we are constantly monitoring the dozens of stars that revolve around Sagittarius-A* with narrow and swift elliptical orbits.

Perhaps the teaching of Chiron, that great and wise Sagittarius, will allow even us poor earthbound scientists to emancipate ourselves, sooner or later, from our abysmal ignorance of these giant celestial objects.

Do Not Wake the Sleeping Dragon

The mass of Sagittarius-A* is certainly enormous, but pales in comparison with the black hole at the centre of NGC-4261, a galaxy in the Virgo constellation. This gigantic object weighs the equivalent of 1.2 billion solar masses.

This is unquestionably an extreme case, but it has now become a widely held conviction that almost every great galaxy has one of these supermassive black holes at its core, with masses ranging from a few million to a billion times that of the Sun. In short, it really does seem that without the presence of these fascinating monsters it would be impossible to create those marvellous objects that we know as galaxies: dynamic configurations of matter that are stable on a timescale measured in billions of years.

The heavyweights among black holes have other characteristics that differentiate them from smaller ones, the result of the evolution of massive stars. They do not have, for example, the extremely disproportionate density of their compact relatives. Giant black holes may have a density lower than that of water, which seemingly renders them less fierce. Their tidal forces, which would be responsible for reducing you to smithereens if you got close to a black hole with a mass of three or four times that of the Sun, are weaker to the point of being almost imperceptible. It would be possible to cross their event horizon without even realising you had done so, at least initially. Despite this apparently docile aspect, however, they are among the most dangerous objects in the cosmos, capable of devastating an entire galaxy. Supermassive black holes are in fact the source of some of the most energetic phenomena in the universe.

For many decades, for example, quasars – a name shortened from *quasi-stellar radio source*, that is, a radio source that resembles a star – had remained an unfathomable mystery. Today they are referred to by the acronym QSO, which stands for quasi-stellar object. Originally discovered in the 1950s, they are the most potent sources of light in the universe. They were initially picked up because they emitted strong radio signals; then, by pointing optical telescopes at the areas where the signals were coming from, very powerful luminous signals were recorded. The active region turned out to be very small, almost pointlike in fact, as if a single star was producing this amazing phenomenon.

But no star could shine with a light a thousand times more powerful than that emitted by the 200 billion stars of the Milky Way. Basically, in those far-off galaxies something mysterious was happening, which had something to do with unusual celestial bodies. Bizarre new phenomena were suggested, but in the end, as ever more complete data were gathered, the conclusion reached proved to be even more sensational: it was the *black stars* that were brighter than anything else. The pointlike bodies emitting so powerfully were at the centre of galaxies in which supermassive black holes were hiding. Often these charming 'dragons' were taking a comfortable nap, just as they do in legendary tales when there is no one to disturb them, but on occasion they made an ostentatious display of their power, 'spitting' flame, light, and every kind of electromagnetic wave for vast distances. In this case we are dealing with so-called active galactic nuclei.

The supermassive black holes found at the heart of most galaxies are frequently very placid, as seems to be the case with Sagittarius-A*, which swallows material, and pulls some stars to pieces, but on the whole behaves in quite a well-mannered and discreet fashion. We've only recently become aware of its existence, and only then because we have been trying so determinedly to look inside the galactic nucleus. Driven by curiosity, we went to have a look at what was happening beneath the blanket of dust that was hiding everything, and we discovered that Sagittarius-A* plays cat and mouse with the stars that are rapidly orbiting around it. Apart from this no one would have noticed anything out of the ordinary.

The nucleus of our galaxy, seen from outside, is not a cause for concern: it does not emit dangerous levels of radiation and does no harm. But we are rather fortunate in this. There are occasions when a galaxy's nucleus enters into a turbulent state of excitement, and at this point things start to turn very nasty indeed. It happens when, around the very core, there is an extremely high density of stars, gas and dust. If, in short, there is a surfeit for the black hole to consume, a kind of feeding frenzy is triggered. It surrounds itself with a huge accretion disc: matter is dismembered and dragged into a vortex around it, a kind of hellish carousel ride on which the very high speeds, collisions and interactions between the fragments of matter create processes that heat everything to temperatures of millions of degrees.

The ionised matter is reduced to elementary components that produce immense magnetic fields which, in their turn, interact with other material. When there are significant accretion discs, we frequently see enormous jets of particles and associated radiation issuing from the poles of the black hole. We are talking about highly energised collimated beams of matter and radiation, emitted by the active nucleus in a direction perpendicular to the plane occupied by the galaxy. The images that we have captured of this are pretty impressive: we can see giant strands of matter that having been generated at the core of the galaxy can extend for tens of thousands of light years. The intense radiation emitted appears in the form of lobes that proliferate from the galaxy, forming protuberances that extend for millions of light years.

The details of this phenomenon are still not entirely clear to us. It is thought that while a section of the ionised matter vanishes into the event horizon and ultimately causes the black hole to grow, a fraction is diverted towards the poles where it undergoes acceleration to frightening speeds. We see in operation in the cosmos hundreds of accelerators that are more powerful than the LHC, producing relativistic jets resembling those studied at CERN but with dimensions similar to those of an entire galaxy.

A small fraction of the active galaxies have their own spectacular jets oriented precisely in the direction of Earth. In this case we can observe a spectrum of electromagnetic radiation amplified by the incredible velocity of the jets, characterised by rapid and violent variations in flow. Historically this type of source became known as a *blazar*, from the first strange object of its kind, BL Lacertae, situated in the Lacerta constellation, with a brightness so dependent on time that many thought it must be a variable star belonging to our Milky Way. With more accurate observations we discovered that it was actually a distant galaxy 90 million light years away. When the origin of this behaviour was related to an active galactic nucleus, the phenomenon joined this wider class.

Quasars, blazars, and active galactic nuclei in general are quite rare phenomena in the universe as a whole, and yet they have been discovered in their hundreds of thousands. Very few are found in dwarf galaxies, whereas they are rather more frequent, found in up to one in five, in those elliptical giants that are the result of the fusion of several galaxies.

It also seems to depend markedly on the age of the galaxy. There is a high proportion of quasars, for instance, in the older galaxies – a sure sign of the fact that active galactic nuclei played a fundamental role in the formation of primeval galaxies. A further proof of this is provided by the fact that the oldest quasar identified dates back to just 700 million years after the Big Bang. In short, they were already present in the first great structures, though the peak of their appearances dates back to around 10 billion years ago, after which the numbers begin to diminish.

The fact seems to be linked to a mechanism of progressive exhaustion of the necessary fuel. The black hole, concentrated into itself, burns up and recycles all the matter that it has managed to extract from its surroundings over billions of years. The very same mechanism, and the extremely strong radiation produced in the process, ends up entirely impoverishing the core of the fuel that it needs. Without new material, the growth of the accretion disc is interrupted and the process is extinguished.

This would explain why so many large galaxies, such as our own for instance, despite playing host to an enormous black hole, do not have active nuclei. There is simply not enough material left. As far as the Milky Way is concerned, we can therefore rest at ease. As long, that is, as it does not enter upon a collision course with Andromeda. If this were to happen, the fusion could transport sufficient material into the nuclear core to reactivate it, and life on any planet in the galaxy could become rather a complicated prospect.

In the end, the role of these 'devouring monsters' situated at the centre of many galaxies appears to be essential to their overall dynamics. The most gigantic black holes are both great destroyers and great creative forces. The frenzied dance they put matter through resembles a spectacular restaging, on a cosmic scale, of that of the Sufi whirling dervishes of Konya. It also recalls the myth of destruction–creation through the dance of Shiva. But above all, by keeping large quantities of stars on this dangerous fairground ride for billions of years, it gives to matter something most precious:

the time that's necessary to produce solar systems, planets, and increasingly complex forms of organisation.

The problem remains of understanding how black holes form with masses a million or a billion times greater than that of the Sun. We know that once a black hole situates itself at the centre of a galaxy, it can grow out of all proportion by gradually swallowing up everything around it. But how does the process start? Perhaps, before the first stars began to shine, the immense clouds of primeval gas aggregated into quasi-stars, objects that are so unstable that rather than evolving into ordinary stars they collapse into black holes. There are some who have even hypothesised that the first black holes came into being less than a second after the Big Bang when the powerful fluctuations in density of the newly born universe could induce the gravitational collapse of enormous portions of matter. The new field that puts at its centre such cumbersome celestial objects as these is still full of mysteries.

Orion's Fine Arrows

While investigating the origins and dynamics of these highly turbulent phenomena, decisive progress is being made in our understanding of other things that until recently had proved to be totally unfathomable, such as where cosmic rays come from.

Since 1921, physicists have been looking for the origin of this shower of charged particles that constantly rain down on our planet from every direction. Some have been recorded as having energies 100 million times greater than those of the LHC, and their origin had until recently remained a mystery.

Everything happened, including in this instance, because different instruments were brought together to observe the same phenomenon, another success for multimessenger astronomy.

It all started with the sound of an alarm given by IceCube, an experiment in Antarctica specialising in the detection of neutrinos coming from deep space. The detection of high-energy neutrinos produced by cosmic sources is a very rare event and requires detectors of colossal dimensions. This is the case with IceCube, an ironic enough name for a detector with the volume of a mountain, a 'cube' with sides of 1 kilometre.

This was undertaken in the Antarctic, not far from the Amundsen–Scott station, in order to exploit the layer of extremely pure and transparent ice that covers the continent. The ice was drilled into, melting it at a hundred different points 100 metres apart and organised on a hexagonal grid. The holes were drilled to a depth of more than 2 kilometres, and into each one sophisticated photon detectors were lowered. When the water around them froze again, the thousands of detectors remained buried in the darkest reaches of the ice. And these electronic and ultra-sensitive eyes began to scrutinise the most absolute darkness in search of the tiniest flashes of light produced by unfortunate neutrinos that expire when they hit a nucleus while crossing through the thick layer of ice.

The high-energy collision produces swarms of charged particles, sometimes accompanied by muons, a kind of much heavier electron that is emitted in the same direction as the neutrinos and finds itself suddenly travelling faster than light in this medium. The only way of escaping their circumstances is to behave like fighter bombers when breaking the sound barrier. But instead of emerging after a thunderous sonic boom, the muons confine themselves to releasing minute flashes of ultraviolet light distributed in a characteristic cone. This was first observed in the 1950s by Pavel Alekseyevich Cherenkov, whose name has been given to the effect.

Thus it is that when a neutrino interacts, the detectors of IceCube record a sequence of characteristic signals that make it possible to measure both its energy and the direction from which it originates. And this information is very significant in that it allows us to work back to the source that has emitted these light, delicate messengers. Cosmic neutrinos fly imperturbably in a straight line, regardless of the distribution of mass and energy they are crossing through, totally insensitive to the magnetic fields occupying galaxies and even intergalactic space. To detect them means to identify the galaxy from which they originated, and to begin to understand the mechanism by which they were generated.

Ever since it first began to collect data, IceCube immediately succeeded in recording some truly spectacular events that astonished everyone: neutrinos possessing frightening energies, one hundred times greater than we can produce using the LHC, the most powerful accelerator in the world. No one could have guessed until then that such high-energy neutrinos existed, wandering throughout the universe, and from this point the challenge was on to understand what kind of supersized cosmic accelerator was capable of producing such particles.

On the 22 September 2017, IceCube's detectors recorded

the interaction of a neutrino of 300 TeV, from which a muon was produced that trailed behind it a spectacular luminous track revealed by hundreds of photosensors. The data was unquestionably clear, and the direction of flight of the neutrino pointed to a far-off galaxy that is well known for being very active in the emission of radiation of various wavelengths. It is around 4 billion light years away, near to the constellation of Orion, the great archer who shines in the northern sky, an everlasting memorial to the giant hunter killed by the hand of Artemis.

The myth tells how Apollo, displeased by the attraction his sister felt for this mortal so skilled at hunting, tricked her into killing him. Zeus, moved to compassion by the tears of his daughter and the inconsolable howling of his faithful hound Sirius, his companion on many hunts, assigns them both places in the most splendid constellations. And still today we can see them above us, hunting together, launching their arrows in the direction of Taurus.

In this case, however, Orion has launched towards us arrows of a different kind, finer and more penetrating than the ones used to bring down deer and wild boar. The neutrinos detected by IceCube come from a galaxy with one of those code-like names to which astronomers are forced to resort, given the myriad of galaxies in the heavens: TXS 0506+056. Physicists, however, are not fond of complication and the galaxy is immediately renamed, in a way that is easier to remember, as 'Texas Source'.

The researchers who deal with gathering data from the experiment send out an alert to all the observatories in the world: 'Scientists of planet Earth, look towards the Texas Source; something is happening up there.' The message is heeded by dozens of observatories that are pointing their instruments in the direction indicated, and this is where the fun starts. In the following days another two devices specialising in the detection of high-energy photons detect gamma rays undoubtedly from the same source. There is no longer any doubt that the Texas Source is putting on a show.

We had known for some time that TXS 0506+056 was a very strange object. We are talking about a truly vast elliptical galaxy dominated by an enormous black hole rotating rapidly on itself. This monster has a gigantic mass, estimated at hundreds of millions if not billions of solar masses, and is adorned with a huge accretion disc and two giant polar jets. One of these is directed towards the Earth, making it a blazar.

From the fearsome accelerations taking place in the Texas Source, gamma rays as well as neutrinos are produced. Gamma rays consist of photons of extremely high energy that activate the instruments of both FERMI and Magic, the two most sensitive observatories, the former being in orbit around the Earth, while the latter has its two telescopes positioned on the island of La Palma in the Canaries.

This was precisely the signal that everyone had dreamed of finding. Such a coincidence can hardly be an accident: if neutrinos are also emitted together with the photons, this is proof that the gargantuan contraption fed by the black hole of the Texas Source actually accelerates protons, just like a titanic version of our LHC.

Thus we have begun to understand one of the biggest mysteries of modern physics, and we have been gifted this knowledge by far-off galaxies fed by giant black holes.

With this we have reached the end of the sixth day; the

first 4 billion years have elapsed and the universe is now populated by a myriad of galaxies. Among them there is a very docile one with a galactic nucleus that has now become calm, and in which something truly momentous is about to happen.

Day Seven: A Swarming of Complex Forms

In the Milky Way everything has now been revolving stably around a central core for billions of years. The turbulent phase of the new galaxy's life, its tempestuous adolescence, is long since over.

Sagittarius-A*, after it has swallowed all the stars, gas and dust that surrounded the original core, has fallen into a long, calm, satiated slumber, like the savage giant Polyphemus in his cave, rendered harmless with wine by Ulysses. The accretion disc of the great black hole, no longer overfed, has shrunk in size, and the relativistic jets with which it radiated the surrounding space, buffeting stars and systems in formation, have gradually disappeared. Even the nearest giant galaxies, the closest cousins of the family that makes up the local group, Andromeda and the Triangulum, have stopped producing spectacularly dangerous fireworks. The gamma rays emitted by the active nuclei of far-off galaxies are relatively harmless. Now, in the calm that has descended, no longer broken by the sequence of catastrophes that has characterised the birth of the galaxy, there is time for the development of increasingly complex organised systems.

More than 9 billion years have passed by the time the last day, the seventh, begins. Something is happening in a minor region compared with the four great structures that make up the immense spiral. Between the great arms of Perseus and Sagittarius, precisely at the point where a smaller arm known as Orion begins, there is a swarming of young stellar formations that find a source of nourishment in the gigantic molecular clouds. In this zone generations of massive stars that have succeeded each other for billions of years have dispersed all the material accumulated in their enormous nuclear furnaces.

Exploding like supernovas, they have scattered vast areas of dust and gas: the molecular clouds. They consist mostly of hydrogen and helium, but there are also traces of all the elements: carbon, nitrogen, oxygen, silicon, and so on up to iron. Some of these great stars, transformed into neutron stars, enriched the clouds with various concentrations of heavier elements, including lead and uranium, when they came into collision with each other.

While they are hot and continuing to expand, a legacy of the great explosions that brought them into being, nothing can bring these immense clouds together. But gradually they begin to cool down and their velocity slows; gravity prevails over the impulse to expand, and around lumps of matter it creates centres of aggregation that become ever larger. Here we find that a great disc of gas and dust is formed, rotating around the centre, where the bulk of its mass thickens, mostly with hydrogen. Within the galaxy a miniature replica of the galaxy is formed: a substantial part of the great cloud collapses under the force exerted by gravity and forms a solar nebula at the centre of which a star is struggling to be born, while all around it a kind of accretion disc is formed in which it is possible to make out other, smaller centres of aggregation distributed in various rings: a protoplanetary disc, as it is called.

Suddenly the Sun will begin to shine, and the first large gas planets will be formed. Then, more slowly and following a bumpier course, the rocky ones of the innermost orbits will aggregate.

One of these will be particularly fortunate. The catastrophic crash with another evolving planet, rather than devastating it for all time and reducing it to thousands of fragments, will produce a huge satellite as a gift that will help to stabilise its orbit for billions of years to come. It will be bombarded, like the others, by a hail of comets and meteorites that will provide it with a wealth of important elements, and all of this, together with the volcanic activity that will accompany it, will play a decisive role in the developments that follow.

The large rocky planet produces a gravitational force powerful enough to surround itself with a gaseous atmosphere, and its molten metal core gives it a magnetic field. These two factors will serve as a protective shield against the many threats lurking in the depths of the cosmos.

It will enter an orbit near the Sun, allowing it to absorb sufficient energy to escape the cosmic cold by which it is surrounded, but not so close that it will heat up to temperatures incompatible with many chemical reactions. The water with which it will largely cover itself will be able to stay liquid for billions of years, and it is precisely within its depths that very special chemical forms will emerge. These are simple structures but equipped with ingenious features allowing them to adapt and develop: chemical systems that incorporate and transform elementary molecules into more complex structures. These are the first forms of life, capable of evolving and reproducing in response to their environment.

The biggest step of all has thus been taken. It has been about a billion years since the formation of the solar system, and on planet Earth the first living organisms are developing. From this moment onwards, slowly but inexorably, the complex chemical forms capable of adapting and changing and colonising ever vaster areas of the planet will rise and fall in succession, with intervals of great success, with booms of this or that species between periods of crisis and mass extinction.

The organisation of living beings offers such advantages that it gives rise to the development of ever more complex forms, from single-cell organisms to plants and animals, and eventually to us. We are almost nearing the end of the story, when in a strange kind of anthropomorphic ape with strong social relationships, natural selection will develop a new tool that will furnish it with a further evolutionary advantage: the ability to imagine, to have a vision of the world and a degree of self-awareness. From then on, this peculiar species of animal will spread to all the corners of the planet and will equip itself with increasingly sophisticated tools to enable it to construct a conception of the world that is progressively more elaborate and accurate, organising around itself its own grand narrative of origins.

The seventh day comes to an end and Genesis finishes, 13.8 billion years after the Big Bang.

The Sun and Its Wandering Stars

At a certain point a portion of the great molecular cloud begins to collapse around an area of higher density. We are in the arm of Orion, a tranquil zone of the galaxy, at a safe distance from the core which, although less turbulent than at first, is still a region in which upheavals periodically occur.

Gravity causes hydrogen, gas and dust to converge towards the place where there is maximum concentration, and everything begins to orbit around this centre of attraction. Through conservation of angular momentum an enormous flat disc forms, inside which the central region of increased density continues to grow. In the eye of this sort of enormous cyclonic vortex, mainly molecular hydrogen is concentrated; at the centre of the disc, crushed by the constantly growing gravitational attraction, a gigantic spherical body comes into being, within which the first thermonuclear reactions are unleashed, and a new star is born.

The dimensions of the Sun are large enough to produce surface temperatures of many thousands of degrees and to carry energy for enormous distances. But it is a dwarf star, and its modest size gives it the advantage of only slowly consuming the ionised and compressed hydrogen of which it is made. The new star will be able to continue to shine for 10 billion years – a substantial period of time, long enough to facilitate the development of a stable system of planets and satellites that in turn will have at their disposal billions of years to accompany the processes of transformation.

The term 'planet' derives from *planetes asteres*, wandering stars, as the Greeks used to call those stars that moved in the night sky in relation to the fixed ones. The wandering stars consisted of the Sun, the Moon, and the five celestial bodies that are visible to the naked eye: Mars, Mercury, Jupiter, Venus and Saturn. These seven planets would soon become associated with some of the principal deities, who will adopt their main characteristics. The ardent and brilliant Mercury that, crossing the sky with such impressive swiftness, will become the agile messenger of the gods; the glittering Mars, with that turbid, blood-red colour it exhibits when low on the horizon, will become the god of war; and so on. These seven will also define the sequence of the days of the week: from Greek they will cross over into Latin, and from there into the Romance languages and almost all the European languages as well, and come down to us today virtually unchanged. The inhabitants of planet Earth have always been, for millennia, so fond of these 'wanderers' that the very passage of time is marked by their names.

But now, as the Sun begins to shine at the centre of the nebula, the various rings of matter that surround it begin in their turn to aggregate at the zones of maximum density. In this way the four gas giants that occupy the outer orbits of the system are formed: Jupiter, Saturn, Uranus and Neptune. All of this happens in a relatively short period of time, around 100,000 years. It will take much longer, tens of millions of years, for the rocky or terrestrial planets to aggregate.

The Sun in its earliest phase of life, like all other stars, puts on quite a show. Its brightness and the radiation that it emits are much more intense than at present. Heated to extremely high temperatures and driven by the winds of charged particles generated by the Sun's magnetic storms, the hydrogen and other lighter components of the original nebula are swept away from its closer orbits. Driven towards the zone occupied by the gas giants, they are captured by them and enclosed within their gargantuan mass. While the protoplanetary nebula begins to become orderly and clear, the inner sanctum of the solar system ends up enriching itself with progressively heavier elements.

The little grains of dust orbiting in the zone closest to the Sun, and which radiation and solar wind fail to sweep away on account of their considerable mass, crash into each other and begin to aggregate into ever larger bodies. When they reach a size of the order of a kilometre, the gravitational attraction that they exert around themselves forms increasingly significant aggregations, eventually producing a myriad of rocky bodies. These are the so-called *planetesimals*, or infinitesimal planets, the seeds from which the planets, satellites and rocky asteroids of our solar system will grow.

Mercury, Venus, Earth and Mars, the rocky planets that are inside the orbit of Jupiter, were born from the aggregation and fusion through chaotic collision of thousands of these small celestial bodies. As the dimensions increase, the heavier parts of the material, typically iron and nickel, will concentrate in the core of the planet in solid form; the pressure due to gravity produces temperatures of thousands of degrees that liquefy the outer layer of the metallic core. Rocks and lighter elements float above it, concentrating in the upper layers; shells of liquid rock encase the metal core, while with complete cooling a rocky solid crust will slowly form on the surface, gradually becoming thicker.

In this way, around 4.5 billion years ago, a highly structured solar system was formed: eight planets, tens of dwarf planets, hundreds of satellites, thousands of celestial bodies of sub-planetary dimensions, and more than 100,000 asteroids. Among these planets there is one that occupies a particularly privileged position, having been the beneficiary of outrageously good fortune.

Thank Goodness for Theia

It sometimes happens, including in our own lives, that good luck comes in the form of bad. Passengers despairing at having missed their flight after being delayed on the way to the airport discover later that they have avoided a plane crash in which there were no survivors. More commonly a defeat, a professional failure that causes a change of job, or a terrible disappointment in love that breaks up a once important relationship can cause us, perhaps years later, to look back and to realise that what seemed like the most miserable period of our life in reality marked a change, the opening of a new door or the opportunity for us to meet the person with whom we will fall in love.

But nothing can remotely compare in this respect to what happened to our planet in the first phase of its existence. Approximately 100 million years have passed since the orbital slot third nearest to the Sun was first occupied by a sizeable rocky planet. We'll call it Gaia, the ancient name for Earth. Like all such bodies, it too formed through the aggregation of planetesimals, and it too underwent periods of great turbulence, characterised by collisions and colossal gravitational perturbations. And just as the worst of this seemed to be over, a disaster turned out to have been waiting to happen.

Another celestial body, smaller than Gaia but still of considerable magnitude, has an orbit that carries it towards an inevitable collision with our own. What happens next is like the nightmare scenario in Lars von Trier's 2011 film *Melancholia*.

The planetoid that is about to smash into us has a mass similar to that of Mars. We'll call it Theia. Powerful tidal forces devastate the two planets even before they collide, and then the even more catastrophic impact arrives. The energy unleashed in the violent collision is such that the two gigantic bodies fuse together for long periods, with shock waves shooting rapidly through them. Then a part of Theia, mixed together with material from Gaia, slips from this fatal embrace and attempts to flee. But it remains trapped forever in the gravitational field of Gaia. And thus it is that our Moon is born. Just as in Greek myth the Titaness Theia, the goddess par excellence, daughter of Uranus and Gaia, gives birth to Selene, 'the resplendent one'.

In its turn Gaia, having absorbed the trauma of the impact and the separation of the Moon, has resumed its spherical form, ultimately magnified its dimensions and become the planet Earth. The hypothesis that the Earth–Moon system had its origin in this catastrophic primeval collision event has found many confirmations in the analysis of lunar rocks collected during various explorations of our satellite. In some oxygen isotopes that have been found inside them there remains a kind of fossil imprint of the fiery original embrace that tied Earth to its satellite.

The Moon serves not only to illuminate our nights, inspire lovers' dreams, and provide inspiration for musicians and poets. This strange satellite, so anomalous with respect to the others that populate the solar system in their hundreds, plays a fundamental role in the stabilisation of our planet's orbit. The Earth–Moon system acts as a sort of gyroscope stabilising their revolution around the Sun.

Earth is the only rocky planet to have such a large satellite: with a diameter of 3,500 kilometres, it is around one quarter the size of Earth. Mercury and Venus do not have satellites, while Phobos and Deimos, the two tiny moons of Mars, are only small ellipsoids with diameters of 22 and 12 kilometres respectively. The three rocky planets that are our companions are exposed to the gravitational disturbance coming from the Sun and the other more massive bodies in the solar system, destabilising the angle between their axis of rotation and the orbital plane. On a temporal scale of millions of years this can cause significant changes: the angle can even alter by tens of degrees, in periods of chaotic change.

The same thing would happen to the Earth if we did not have the Moon, with its weight and proximity, to attenuate the disturbances that would modify our axis of rotation. The angle that the Earth makes with the orbital plane is stabilised by the presence of the Moon, so that there is a variability of the order of just one degree. If the inclination of the Earth with respect to the Sun remains fixed, then it is possible to define climatic regions that remain stable over long periods of time – something that tends to favour the development of the very slow processes that produce complex systems. If someone were to address the Moon again like the nomadic shepherd from the Asian subcontinent – 'What are you doing there, in the night sky? Tell me, silent Moon, what your secret is...' – he might receive the reply, maybe less poetic but surely original and surprising: 'Without me there would be no seasons, and perhaps even no life on Earth; not even nomadic shepherds asking me questions as they gaze up at my face.' Earth's devastation by Theia turned out to be an instance of tremendously good luck.

It is not the only one. The other major stroke of luck

happened as a result of having the colossal Jupiter nearby. The great gas giant, the heavyweight champion of the solar system, has a diameter of 143,000 kilometres and weighs in at three times heavier than Earth. It is so abnormal that even today there is debate as to whether it is really a planet at all or a small so-called brown dwarf instead. When the initial mass of the gas sphere is not sufficiently large, the pressure and temperature of the nucleus cannot manage to trigger thermonuclear fusions; yet the body is hot enough to radiate a substantial amount of energy. The failed star becomes a tepid astral body that irradiates at much lower temperatures; its light is not bright like blue, white or yellow light, but tends towards a dull redness. Hence the term brown dwarf.

Jupiter, the failed star, nevertheless has such an imposing mass that it has in large part conditioned the development of the solar system. One of the first to be formed, with its dominant force of gravity it has prevented the formation of a rocky planet in the so-called asteroid belt, an extensive region between Jupiter and Mars. They have been pushed in great quantities towards outer space, and it has prevented any others from consolidating into a massive body. In the belt there are still thousands of pieces of rocky detritus, vestiges of those bodies that the attraction of their bullying neighbour has catastrophically affected, forcing them to collide continuously every time they attempt to organise themselves into a planet. The failed formation of a fifth rocky planet has left behind more material, structured into planetesimals, for the formation of the inner planets, including Earth. Our planet has thus been able to acquire the dimensions necessary to succeed in permanently retaining its atmosphere.

The good giant, Jupiter, along with Saturn armoured

with its rings, act as sentinels guarding the inner planets. With their huge mass they divert towards themselves and incorporate potentially dangerous asteroids and comets. Like gigantic bodyguards they shield us from the risk of being drawn into highly perilous encounters with other objects. They are not always successful in doing so: 65 million years ago, a meteorite rich in iridium with a diameter of 10 kilometres managed to reach our planet. But events as destructive as this, thanks to their presence, have become extremely rare occurrences for us.

Jupiter's great shield protects us from catastrophic events that could threaten the survival of the delicate forms of life that have developed on Earth. For this we owe a debt of gratitude to that great planet Jupiter, the mediator, the peacemaker able to moderate conficts between the gods, that the Greeks were right to identify with Zeus himself.

The Cradle of Complexity

The secret of the Earth is hidden in its innermost depths. Above its solid nucleus and a shell of molten metal floats a thick layer of liquid rock. Ever since the original formation of the planet, iron and other heavy metals have been differentiated from lighter components. The heavy metals consolidated in the innermost strata, while the others aggregated to form a thick rocky outer layer. The heat generated by gravitational contraction melted the inner part, while with cooling a thin rocky crust was created that floats on the sea of molten rock. Processes of radioactive decay of unstable isotopes feed, with their energy, the heat of the core and help to maintain the high temperatures for billions of years.

The great rocky plates of the crust are constantly moving, pushed by the energy of enormous convective cells that form in the underlying mantle of melted rock. From the titanic crashes that ensue deformations occur, creating mountains and deep valleys that will be filled with the waters of the oceans. From the fissures that are created an incandescent magma that roars beneath the crust now surfaces. The god of fire, Vulcan, the blacksmith in his echoing underground workshop, works unceasingly to build the marvellous landscapes of the Earth.

In its initial formative phase the Earth will be overwhelmed by volcanic phenomena of awesome power and intensity. This turbulent volcanism will bring to the surface a continuous flow of chemical substances dissolved in gases and in the molten rock that emerges to form a new crust. Slowly an atmosphere will form, composed mainly of water vapour, nitrogen and carbon dioxide, which the gravitational field of the great rocky planet will be able to retain.

Water was already present in the dust of the protoplanetary cloud, and its molecules will have mixed with those that have formed the rocks of the terrestrial mantle. A large part of this will be lost through evaporation during the hottest phases in the formation of the planet, but the continuous volcanic eruptions will bring it back to the surface in the form of vapour. Most of the water on the planet comes from the incessant flow of asteroids and comets that continue to assail it. The continuous bombardment by carbonaceous meteorites, rich in water, and the cosmic icebergs that are the comets, enrich the Earth with this new element.

By the time the universe reaches its ten billionth anniversary, large oceans cover the majority of the surface of our planet. The volcanic eruptions feed the high concentration of carbon dioxide in the atmosphere, the effect of which will be to maintain most of the water in the oceans in a liquid state for long periods of time.

Phenomena analogous to those that have laid siege to the Earth have brought water to many other bodies in the solar system. It is present in the form of vapour in gas giants such as Jupiter, Saturn and Uranus, and in the clouds that cover Venus. There is ice in the polar caps of Mars, and Europa – the smallest of Jupiter's satellites, discovered by Galileo – is covered by an immense frozen ocean more than 100 kilometres deep, beneath the surface of which we can assume that there is an abundance of liquid water. Titan, the great satellite of Saturn, contains more water than the Earth, but here too, as far as we know, it is in the form of ice. There is probably liquid water on Enceladus, another of the satellites of the great ringed giant.

Earth's incandescent core affords us another gift that will turn out to be very important for the planet's development. The concentric layers of molten iron that rotate at different speeds around the solid inner core drag charged particles along with them and produce an enormous circular current, the source of the thin magnetic field that surrounds the Earth. This invisible structure, directing the charged particles towards the poles, protects the planet from the potentially destructive effects of the cosmic radiation that can easily fracture the bonds that hold together the most complex chemical structures. With this we now have all the key ingredients in place to initiate a sequence of events that will very directly concern us.

Carbon, hydrogen, oxygen, nitrogen, phosphorus and

sulphur are the building blocks for the main organic molecules; they are present almost everywhere in the universe and were certainly also abundant in the environment of the primeval Earth. Starting from these elements, the precursors of the main biomolecules that we find in living beings can be produced in the depths of the oceans near underwater volcanoes or hydrothermal vents; and it is in particular environments such as these, where water heated to high temperatures is enriched with salts and mixed with gases of various kinds, that we need to look to see the emergence of the first biological structures. There we may find chemical reactions that have transformed carbon monoxide, ammonia and formaldehyde into amino acids, lipids, polysaccharides and nucleic acids, and that were able to operate for a sufficient length of time to build the earliest proteins and to organise the information in the most primitive forms of DNA.

We also need to consider the hypothesis that bacteria or other very simple living organisms, capable of surviving at extreme temperatures, may have been transported to Earth via the asteroids and comets that bombarded it relentlessly for the first million years of its existence. Embedded in the comets' rocky debris or in the dust mixed with ice, early forms of life originating elsewhere and projected through space by colossal collisions or by gigantic eruptions may have disseminated living material throughout the entire solar system. If the first forms of life had indeed come from outer space, they would surely have found a favourable habitat on Earth.

What's certain is that 3.5 billion years ago, beneath the mantle provided by the water of the ocean, protected from bombardment by ultraviolet rays, the first elementary biological structures begin to evolve: cyanobacteria, extremely small algae whose development initiates another epochal change. They are single-celled organisms that organise themselves into tiny strands, with dimensions smaller than a thousandth of a millimetre, and are *prokaryotes*, which is to say, their genetic material floats freely inside the cell, without the protection of any membrane.

The cyanobacteria are capable of capturing light and converting it into energy – the process known as photosynthesis – and they will perfect this mechanism, adapting it to the diverse environments in which they will develop their colonies.

The biochemical reaction that starting from carbon dioxide and solar light leads to the synthesis of sugars and to the release of oxygen brings a radical change to the Earth's atmosphere. At first the oxygen produced by the algae is absorbed by the iron that was plentiful at the bottom of the oceans. But when the population of cyanobacteria begins to grow excessively, that part of the oxygen that cannot be absorbed by the iron emerges from the water and causes widespread devastation. The composition of the Earth's atmosphere changes radically, eventually becoming ever more toxic for all those organisms that are not able to adapt to the changing environmental conditions. What follows is the first mass extinction of an enormous variety of primitive life forms. But this makes way for the headlong development of new species.

Approximately 2.4 billion years ago the Earth possessed an atmosphere permanently containing a small percentage of oxygen. For we humans this air would not yet be breathable, but progress in that direction was now unstoppable.

Living organisms, the inheritors of the first prokaryotes, developed a protective nucleus for storing genetic material, and the evolutionary advantage this gave determined the success of the eukaryotes. The newly established atmosphere with its oxygen content seems to have favoured the development of the first multicellular organisms, which recent discoveries have allowed us to date to approximately 2 billion years ago. From here on there is a proliferation of a variety of ever more complex biological forms which go through various stages of crisis and of expansion, and by modifying themselves manage to survive terrible mass extinctions.

A veritable phantasmagoria of new living beings occurred around 500 million years ago, when the Earth experienced enormously high temperatures. The levels of carbon dioxide in the Cambrian period reached twenty times those of today, with an average temperature 10 degrees higher. The result was a real explosion of life, with the appearance of a whole variegated spectrum of vegetable forms and of the first vertebrates, fish, and later on the great reptiles.

A new cataclysm radically alters this scenario. With the impact of a large meteorite, 65 million years ago, the climate undergoes a profound transformation due to the dust clouds raised by the collision. A sudden glacial drop in temperature grips the Earth, causing the mass extinction of the great dinosaurs, while affording an unexpected opportunity for smaller mammals to survive and benefit from the extinction by occupying all the ecological niches that had been made vacant.

From one of these, a few million years ago, in a zone made up of gorges and savannah in the Horn of Africa, a population of primates will differentiate itself from the species that preceded it through its remarkable social attitude and its until now untried capacity to invent, make and utilise tools. This spark of self-awareness, which is translated into projects, visions and the making of implements, will constitute an outstanding evolutionary advantage for the anthropomorphic apes. Successive generations of the first hominids will go on to quickly colonise all the planet's habitats, swiftly adapting to diverse environmental conditions.

And so here we are, we have arrived. In the blink of an eye, the story has reached all the way down to us.

Exoplanets

The idea that the universe could contain a multitude of inhabited worlds dates back to the pre-Socratic philosophers of Ionia. The original insight has been attributed to Anaximander of Miletus, the pupil of the brilliant Thales, who was also the first to propose the revolutionary idea that the Earth floats in space, without falling and without any kind of support.

The concept of *infinite worlds* will be taken up first by the Pythagoreans and then, with exceptional lucidity, by Epicurus and his followers in the Roman world, starting with Lucretius. For centuries the idea will be stifled by the dominance of Aristotelian thought, tentatively re-emerging with William of Occam before eventually exploding in the Renaissance with Nicholas of Cusa and Giordano Bruno. It was the philosopher Bruno who with admirable conviction disseminated throughout Europe the idea that there were *innumerable Suns and Earths*, and it was probably the activity of publicly advocating such dangerous ideas, so far outside the narrow confines of the specialists, that brought him to his tragic end in 1600, burnt alive as a heretic in the Campo de' Fiori in Rome.

Today science has confirmed the intuitions of this line of brave thinkers, and yet we are still struggling to find answers to the simplest question: does intelligent life exist somewhere out there? The law of large numbers suggests that this may be the case – that it is highly probable in fact – but the evidence gathered to date is not sufficient to arrive at any conclusion.

The situation has been rapidly evolving for the last thirty years, due in particular to the huge progress made into research on *exoplanets*. This is the term used to refer to extrasolar planets that orbit around stars different from our own Sun. Until very recently it was thought that the fraction of stars that host planets was very small. In the last few years, ever since techniques were developed to detect them, a month does not go by without a public announcement of some new, groundbreaking observation. At present some 4,100 exoplanets have been discovered.

The earliest efforts to detect them date back to the 1940s, but it was customary at the time to use techniques such as astrometry that were relatively crude. The laws of physics require that, in the presence of a planet, the mother star also makes a small rotation around the system's centre of mass. The more massive the planet, the greater the periodic shift of the star. So they were looking for a small periodic shift in the position of the mother star, but the results were disappointing.

The first real surprises came when they began to use

radial velocity instead, which employs the same principle but utilises spectroscopic measurements that make for greater accuracy. The spectrum of the star's light emission is analysed, and over time the lines corresponding to the various frequencies are checked. If the star exhibits an orbital movement due to the presence of a planet, a measurement is made of the small periodic variation in the frequency of its light emission resulting from the Doppler effect.

It was thanks to this innovative technique that the first extrasolar planets were discovered in the 1990s. And we are talking about enormous celestial bodies, similar in size to our Jupiter. These hot, mostly gaseous giants gravitated very close to their mother stars and therefore had surface temperatures that were truly astonishing.

The field has received an extraordinary boost since the method of using transits was developed, thanks to which it is possible to keep under observation hundreds of thousands of stars at the same time. It's a technique based on precision photometry, whereby the brightness of the star is carefully monitored and the slightest attenuation of the light produced by the planet when it transits in front of it is measured. In this case, too, it is an indispensable requirement that the disturbance should have a periodic character. The characteristic shape of the disturbance allows the dimensions of the planet to be measured, and this information, combined with the measurement of radial velocity that gives the mass, allows us to determine its density.

The sensitivity reached by some of the most modern instruments is such that the field of observation can be extended to thousands of light years, and it has become possible to distinguish planets that are even smaller than Mercury.

In this way, for some years now, research into new 'worlds' has produced tremendous results. It is now clear that an enormous number of the stars in our galaxy are surrounded by planets. It is only a matter of time before we discover some with atmospheres, and therefore with the potential to have developed forms of life similar to our own.

If an exoplanet is surrounded by an atmosphere, the light of the mother star will reach us after it has passed through its upper layers. This passage slightly alters some of the properties through which it is possible to recover essential information. With prolonged observations we will soon be able to establish not only whether some planets have an atmosphere, but whether that atmosphere contains water, carbon dioxide or methane. This will obviously not be enough to be certain that these are signs of life, perhaps with affinities to life on our own planet. But what the numbers and probability are telling us is extremely striking.

If we consider that in every galaxy there are something like 100 billion stars, we must assume also that there is an enormous number of rocky planets out there. Even if we exclude those that orbit in uninhabitable zones, there will be very many that are compatible with the ingredients of life, that is to say planets that are capable of having water in a liquid state.

As we have already seen, this does not by itself guarantee conditions that are favourable to the development of precarious and complex biological structures. The mass of the planet plays an important role, for it needs to have sufficient magnitude to retain an atmosphere subject to its gravity. There would then need to be a magnetic field to protect the planet from cosmic radiation; and finally it would be extremely useful to be able to count on a stable orbit and to be located in one of those galactic regions that exist far from the prospect of catastrophic events. But above all there would need to have been sufficient time, with the required conditions for stability having lasted for billions of years.

Some time ago a NASA probe named after the great German astronomer, Kepler, made a public announcement boasting that no less than 1,284 new extrasolar planets had been discovered. A group of Belgian astronomers studying data from the observatory in La Silla in Chile have identified Trappist-1, a solar mini-system orbiting a red dwarf, a small sun found just 39.5 light years away from us, belonging to the Aquarius constellation. It contains seven rocky planets, a few of them remarkably similar to our Earth – with three of them found in the so-called habitable zone, which is to say at such a distance from the mother star as to allow for temperatures similar to our own. If they have water, this would be able to form into lakes and oceans like those that are so widely distributed on Earth. Now that we know where to look, we will be able to understand all their salient characteristics better, and perhaps to gauge whether one or more of these planets has an atmosphere.

Based on what we already know, Trappist-1 is much too young to contain forms of life, given that the small solar system is only 400 million years old. But we are definitely at the beginning of what should be a long series of discoveries. The countdown has already started. In a few years' time, when we will have at our disposal the first unequivocal data and have dispelled all final doubts about the existence of another inhabited world, we will be faced with a double challenge: on the one hand coming to terms with and absorbing such a cultural shock; on the other – and why not, despite the distances involved – developing technology adapted to making contact with or even to visiting the new worlds that we will have discovered. Once again, we are witnessing how science can proceed by giant steps, and change at a stroke paradigms that once seemed immutable.

But let's return now to our story of origins, which reaches its conclusion when 13.8 billion years have passed. The seventh day ends precisely at the moment when our distant ancestor rises to his feet, begins to tell the story, and the others form a circle in rapt suspense, eager to listen.

The Human Factor

No one will ever know when it happened exactly, nor is it possible to know who was the first. There is no hope of reconstructing the language they used, much less the message communicated to this small group. Perhaps it was to celebrate a moment of collective euphoria and joy, or to seek consolation after some terrible loss.

What we know for certain is that someone, at some point in our history, began to tell a story. No doubt it was an individual more extrovert than the others, affected perhaps by some psychopathology – or just some more than usually restless soul who happened to be able to string words together in an arresting way. We can only imagine the scene: within a dimly lit cavern, a family clan of ten to fifteen individuals sitting around the one who has discovered the power of fascinating others, of linking them all together with a miraculous string of words. A chain of expressions used in a new context, freed from utilitarian function, finding the space to become song, poetry, collective knowledge. Ritual words that take on a profound symbolic value, and captivate those listening to them.

The Construction of the Symbolic

Various finds and discoveries that have followed one after another in recent decades have led us to conclude that the first symbolic universe was developed by the Neanderthals. We are talking about a species present in Europe hundreds of thousands of years before the arrival of *Homo sapiens*.

Both species seem to derive from a common ancestor, *Homo heidelbergensis*, who had evolved in turn in Africa from *Homo erectus*, more than a million years ago. After colonising the African continent, this species would have spread to Europe, and perhaps to Asia, during the interglacial period around 600,000 years ago. Sapiens differentiate themselves from the Heidelbergs that remained in Africa; the Neanderthals will descend from those that colonised Europe. The two species, evolving in environments and contexts completely different from each other, develop distinct characteristics – but from a genetic point of view they remain very closely linked. We are talking about close relatives: if not exactly brothers then cousins at the very least.

The physical characteristics of the Neanderthals have almost certainly contributed to a certain prejudice against them. More robust and heavyset than the slimline Sapiens, they have always been portrayed by their relatives as more primitive and less developed. In reality these physical characteristics were the result of an extraordinary adaptation to a very difficult environment.

The Europe in which the Neanderthals lived for hundreds of thousands of years had a very harsh climate in which brief periods of warmth alternated with extremely long glacial eras that would have severely tested the capacity for survival of any species that endured them. Lack of sunlight will cause the development of a genetic mutation in the Neanderthals that will cause them to have white skin, much lighter than that of their ancestors and of Sapiens as well, as would have been evident when we first arrived from Africa and encountered each other for the first time. Many have chestnut, blond or red hair, and light eyes; all have a sturdy physique, solid bones, and well-developed muscles, characteristics crucial for dealing with an inflexible climate and for surviving in such hostile terrain. Their cranial capacity is greater than that of Sapiens: they have a larger brain than ours, though their head is ovoid in shape, resembling a rugby ball, with a low, protruding forehead and a pronounced occipital bone. They have large noses, monobrows, and marked prognathism, or jutting jaws.

Basically, the physical appearance of the Neanderthals contrasts completely with the canon of beauty that we have developed in our own image. But if we were to encounter one now on the underground, dressed for work in a suit and tie, we would not be overly astonished by their appearance. Among the infinite variety of human individuals, similar features to those of this ancient species can be found. And despite their 'primitive' appearance and reputation, it seems that it was these very cousins of ours, behind their rough exterior, who were able to develop one of the most powerful instruments of survival: a symbolic universe.

The Neanderthals are muscle-bound athletes with a protein-rich diet, the only kind that would allow them to survive in the freezing climate of an ice-bound Europe. For shelter and clothing they use animal skins that they flay and dress with masterly skill; they have strong hands with which they fashion sophisticated tools and utensils from stone and wood. They are expert at turning flint into pointed, bladed tools, using a technique that will come to be known as Mousterian, and that will spread throughout Europe the products of their extraordinary craftsmanship: sharpened tips, discs, blades, scrapers, and beautiful bifacials or double-edged blades. Many of these, in the form of spear-tips or blades, will be fixed with bitumen into lengths of wood, turning them into lethal weapons such as long spears.

Neanderthals are omnivorous, but 50% of their diet consisted of meat. Finding a large carcase, they would become opportunistic feeders on carrion. But above all they are consummate hunters of live prey. They use spears with hardened tips tempered in fire, and lances more than two metres in length, and with these weapons they are able to hunt the largest animals, including bears and elephants.

In order to organise a large hunting party you must have a strategy, a plan of attack devised with the other members of the group, using sophisticated forms of communication and well-defined hierarchies. You need groups that are willing to shout and make a noise to 'beat' and herd wild animals to a prearranged place, or towards a trap where the strongest and bravest can pounce, or give the *coup de grâce* without taking too great a risk. It is likely that the whole clan would be involved in the hunt, despite the fact that it was fraught with danger. Members of the hunting party would often emerge from it with terrible wounds, from injuries that have been reconstructed from the many fractures discovered in finds of their bones. Evidence of seriously injured individuals surviving into what was advanced age for the period also suggests that the groups had the ability to treat their wounds and care for these individuals, something that would not have been possible without help from younger members and the support of the community as a whole.

With social organisation as developed as this, it is hardly surprising that the Neanderthals also enjoyed a complex cultural life. The archaeological discoveries do, however, provide some particular surprises: there is evidence of the fact that they buried their dead in a foetal position and painted their bodies red. Ornaments have been found painted with ochre, with feathers attached; necklaces have come to light made of deer teeth or the talons of eagles.

The widespread use of ochre was especially significant, since red is the colour of blood, and blood connects life to death. If bodies were buried in a foetal position and painted red, perhaps this was to make a symbolic connection between death and rebirth. This is a crucial clue. A society formed by small groups under relentless pressure from the demands of survival dedicates precious time and energy to the ritual preparation and burial of their dead. Evidently this civilisation gives value, almost more than to the necessity of food, to this symbolic universe; to the extent of deeming indispensable that set of rites and ceremonies that feeds and gives substance to their outlook on the world.

Other findings would seem to reinforce this hypothesis. In a deep cave, hundreds of metres from the entrance, imposing stone circles made by using fragments of stalactites have been found. Who compelled these groups to penetrate such distances, in the pitch dark, along winding passages, into the viscera of the Earth? Why labour under the burden of carrying and splitting in a preselected place heavy stones weighing tens of kilos? And why expend energy arranging them in a circle? This was undoubtedly a practice with a specific meaning; the circular structure has a ritual function that we may never discover, but that was regarded as fundamental enough to justify the time and effort devoted to building it. Something similar can be imagined for objects with less impressive dimensions but equally intriguing purpose: bones incised with geometrical designs, or fashioned into small flutes; some bifacials or double-edged blades carved in rock crystal or other precious stone, never used for practical, everyday purposes and perhaps connected with irredeemably lost ritual ceremonies.

Any doubts about the existence of the symbolic universe of the Neanderthals evaporated when it became possible to accurately date some cave paintings discovered in Spain. A dozen or so examples survive in the interior of three caverns, and they have been found to be more than 65,000 years old; from a time, that is, 20,000 years before the appearance of *Homo sapiens* on European soil. To complete the revelation, in the Cueva de los Aviones, a cave in southeast Spain, researchers have uncovered many perforated and decorated seashells, some of them containing traces of red, yellow and black pigments from 115,000 years ago. They are, perhaps, vessels used to prepare the colours to make the paintings on the wall – of groups of animals, dots, geometrical figures and handprints – in ochre and black.

We do not know what these signs, pictures and graffiti represented to the people who made them. There are symbols, a ladder, animals and hunting scenes. They have been drawn with mastery, with confident technique and a steady hand. When looking at the cave paintings of our distant ancestors there is a tendency to interpret them naturalistically – even those marvellous images produced by Sapiens, tens of thousands of years later. I am thinking of the 18,000-year-old paintings in Altamira and Lascaux. They are of a long series of animals, some men and hunting scenes. But do we really believe that it was worth descending into pitch-black caves, to illuminate them with the flickering light of torches or carefully placed fires, to search for the pigments and to carefully mix them, and to practise for years, simply in order to depict scenes of everyday life?

Behind every individual hand used in making those paintings there is a school, and evidence of great discipline and rigorous selections. Only very few, gifted individuals could be exempted, at least in part, from the hard work of basic survival in order to pursue such activities instead. We have to imagine among the Sapiens, and before that among the Neanderthals, certain great masters who passed on their technical know-how, who chose the most promising among their pupils, those who would be entrusted with the legacy of such precious techniques. To argue that these paintings served to instruct the young in hunting techniques is equivalent to maintaining that the index finger of God the Creator that brushes against Adam's on the ceiling of the Sistine Chapel represents a typical Jewish greeting. Behind the details in these paintings there is a symbolic universe, the architrave of an entire society that was seeking to celebrate and perpetuate itself.

We will never discover the significance the paintings had for the Neanderthals, but we do know that in their eyes they had immense value. The rites and ceremonies performed in these caves were probably of vital importance in providing their society with cohesion. The prejudice that *Homo sapiens* was able to supplant the Neanderthals because we possessed a richer language, a more articulated social structure and a more developed symbolic universe is turning out to be utterly misleading.

The emergence of symbolic thinking signals one of the fundamental stages of human evolution. Today we realise that the more sophisticated cognitive capacity that this development indicates is not a prerogative of *Homo sapiens*, and that it originated much earlier and was something shared with the Neanderthals. Perhaps, in order to establish its actual origins, we will need to revisit remote sources, concentrating our research on the first Neanderthals, or travel further back still, to the ancestor we have in common with them.

What's certain is that the construction of this great narrative of origins, so closely connected to the process that has made us human, has its roots in the dark recesses of time.

In the Beginning Was Thauma, or Wonder

In his *Theaetetus*, Plato remarks to Socrates: 'This pathos is proper to the philosopher: it is the *thaumazein*. And philosophy has no other point of departure than this.' The word, which contains the root *thauma*, the same that appears in thaumaturgy, has often been translated as 'wonder'. Philosophy is born out of amazement mixed with the curiosity that arises from facing something inexplicable that fascinates and transcends us. Aristotle writes explicitly that, beginning by asking the simplest questions, humanity has come to wonder about ever more complex things, ending up by investigating the Moon, the Sun and the stars, and by asking how the very universe itself came into being.

The sense of wonder we get when looking at a star-studded sky is a powerful one, even today an intense and even emotional experience, connecting us perhaps with an echo of that ancient amazement shared by thousands of generations before us. But perhaps too this feeling is not enough to understand the origin of this deep-seated, urgent, primordial, almost innate need to seek an answer to the big questions.

The theme was reprised by Emanuele Severino, a contemporary philosopher, who insisted emphatically on translating *thauma* as 'wonder mixed with anguish'. In this way we recover the original significance of the word, and the knowledge would act as 'an antidote to the terror provoked by the annihilating event that comes out of nowhere'.

In fact, the term was also used in this way by Homer, who speaks of *thauma* when describing Polyphemus, the one-eyed giant who dismembers and devours the unfortunate companions of Ulysses. In this case the link with anguish, implicit in the word, is more self-evident. The very sight of the mythical Cyclops, a creature of colossal size, causes both amazement and terror. The giant, symbol of the untamed force of nature, provokes a sense of wonder at his incredible strength, and at the same time a profound anguish due to our sense of vulnerability and irrelevance. The unleashed forces of nature, an erupting volcano or a terrible hurricane, simultaneously both fascinate and terrify, because they smash our world to pieces or engulf us in an instant. In this bigger picture the role played by such small, fragile beings as we are, continually exposed to suffering and to death, is completely insignificant.

This is where the narrative, the explanation – whether mythical or religious, philosophical or scientific – at that precise moment comforts and reassures us, giving order to an uncontrollable sequence of events and in this way protecting us from anguish and terror. This narrative, in which everyone has a role and everyone plays his part, gives meaning to the great cycle of existence. We are reassured because we feel protected, and our fear of death fades. We remain fully conscious that for us, everything will come to an end – and will do so rapidly compared to the great temporal cycles of evolution in the material structures that surround us. But knowing that the whole obeys an order described in our narratives, we are reassured.

For millions of years, humanity had to come to terms on a daily basis with the harshness of life. Only a few decades ago, and even then for only a part of the world's population, did this experience of extreme fragility and total precariousness subside. But in the depths of our soul we still feel that ancestral anguish. We are all like Leo, the child protagonist of *Melancholia*, who when faced with the inescapable catastrophe that is about to overwhelm the Earth, seeks protection and consolation. He needs someone to say: don't be afraid, nothing will happen to you. He finds this in the person of his aunt Justine, someone who until now has suffered from deep depression, but who when the moment of crisis arrives, when all the 'normal' people are losing their heads, behaves with the most lucidity, and finds resilience enough to maintain her humanity. The small tent in which she seeks refuge with the child will not protect her from the imminent disaster, but until the last moment before the collision, within the warm embrace of his aunt's arms, listening to her calmly voiced story, the child feels safe.

Art, beauty, philosophy, religion, science, in a word

culture, are on one level our magic tent – and we have needed it, desperately, since time immemorial. In all probability they were born at the same time, they are different modalities in which symbolic thought is articulated. It is not difficult to imagine that rhythms and assonances in the use of words would have facilitated the mnemonic transmission of the story of origins, and that this is how song and poetry developed; that something similar may have occurred with signs and symbols depicted on the walls of caves, with increasingly sophisticated formal perfection; or that in the rites and the ceremonies that accompanied moments of mourning, regular sounds would accompany the rhythmic movement of the body or the song of a wise man or shaman. Science is part of this story; it is no accident that *episteme* and *techne* go together, knowledge and the capacity to produce utensils, artefacts, machines.

It was no accident either for the Greeks that *techne*, the root of 'technique', also indicates the common ground between the artisanal and the artistic, and this is why when flint bifacials are produced, the technical requirements of having at one's disposal a sharp and easy-to-handle cutting tool are interlaced with the aesthetic ones of producing something symmetrical, fine, perfectly balanced – in a word, beautiful, like an art object.

These exigencies seem to have constituted something irrepressible for all the human groups that have trod the Earth for millennia. Even the most isolated of remote tribes, found from time to time in some forest in Borneo or the Amazon, have developed their own rites, a specific form of artistic expression and their own symbolic universe, all backed by an overarching story of their origins. Without such narratives, not only would it not be possible to build great civilisations, but even the most elementary social structures would fail to survive. This is the reason why all human groups on our planet are characterised by strong cultural traits.

The Power of Imagination

Culture, the awareness of one's own deepest roots, is a kind of superpower that guarantees a good chance of survival even in the most extreme conditions. Imagine for a moment two primitive social groups, two small clans of Neanderthals that live isolated from each other in the frozen Europe of that era. Now suppose that by chance one of these groups develops their own distinct vision of the world, cultivated and perpetuated over generations through rituals and ceremonies, and perhaps visually represented in cave paintings, while the other group fails to do so, evolving that is without developing any sophisticated form of culture. Now let's suppose that a disaster strikes both groups: a flood or a period of cold even more extreme than usual, or an attack by ferocious beasts that leaves only a solitary living survivor. This last man standing, in the case of both groups, will have to overcome a thousand dangers, face every kind of privation, perhaps migrate to other zones and even evade the hostility of humans. Which of the two will show the most resilience? Who will have the best chance of surviving?

A creation story, a narrative of origins, gives you the strength to get up when you are knocked down, the motivation to endure the most desperate circumstances. Clinging to the blanket that gives us protection and an identity, we find the strength to resist and to carry on. To be able to place ourselves and the others in our clan in a long chain of events that began in a distant past gives us the opportunity to imagine a future. Whoever has this knowledge can place in a wider framework the terrible vicissitudes of the present, giving sense to suffering, helping us to overcome even the most terrible tragedies.

And that is why we are still here, thousands of generations later, to give value to art, philosophy, science. Because we are the inheritors of this natural selection. Those individuals and groups most equipped to develop a symbolic universe have enjoyed a significant evolutionary advantage. And we are their descendants.

We should not find the power of the symbolic and the strength of the imagination surprising. The condition of being social animals is something more profound and intrinsic than the mere fact that we live in organised groups of individuals.

In the last few years, throughout the world, very ambitious projects have been initiated to study the functioning of the human brain. They are well financed and resourced multidisciplinary initiatives, employing thousands of scientists. In many cases, in order to understand in detail some of the basic mechanisms, nets of electronically simulated neurons and their interactions have been produced. All of this is very useful for understanding certain dynamics of the functioning of the brain. So why do the very same neuroscientists that welcomed their development tell us that it would make no sense to expand these elementary structures in order to create an artificial brain?

It is not simply a question of overcoming some major technical difficulties: our cranium hosts almost 90 billion neurons, each of which is capable of establishing up to 10,000 synapses with its neighbours. The question is a deeper one. Even if we were able to construct an electronic device as complex as this, capable of technically reproducing the structure of our brain, it would still fail to be a human one. Still missing from the faithful copy would be an essential ingredient, immeasurably more difficult to reproduce in electronic form. What would be missing is the interaction with other human brains, mediated by language, the body, and emotional relationships. In other words, one becomes human through the perceptions of others, in their eyes and in emotional exchanges with them, by interacting with other humans in relation with us in the social group.

The pliant brain of a newborn infant is shaped in relations with the world mediated by the adults who take care of her, starting with the maternal gaze. A baby looking into the eyes of the person who is feeding her modifies her synapses on the basis of the reactions that occur in their relationship. The thing that we call the human brain is born from the interaction between this plastic system, capable of adapting and being shaped by the stimuli that come from outside it, and by a set of relationships that are established with the rest of the social group: relationships nourished by hopes and desires, beginning even before the embryo is established in the body of the mother. The new being is in dialogue with the aspirations of the parents, which precede birth, and comes into contact with the past and the humans who have come before it. She is projected towards the future through the phantasmagoria that builds the small social group surrounding the figure of the new arrival: grandparents or parents and other family members discern resemblances that link back to ancestral tales in which old fears and new expectations arise. No electronic device could possibly reproduce all of this.

In order to confirm the importance of what we are attempting to describe, we need only think of those instances of very young children lost or abandoned in wild places and raised in the company of animals. They have brains that to start with are structurally identical to those of their contemporaries, but which have failed to become fully human because of the lack of formative human contact. No amount of subsequent contact and attempted rehabilitation will be able to fill the gap created by the missing early interaction.

When imagination and narrative are cultivated within a group, they become powerful tools of survival. Whoever listens and imagines the experiences of others acquires in this way real knowledge. Narrative condenses the lessons accumulated over a long sequence of preceding generations, allowing us to experience and to understand – allowing us to live, in effect, a thousand lives. Imagination allows us to experience emotions and fears, sorrows and dangers, and the values of the group; the rules that help to preserve it and the rules that preserve and govern its development are reiterated and memorised through the generations.

Imagination, as developed and encouraged in groups that are socially and culturally more advanced, is the single most effective weapon that we have ever managed to develop. Science also originates with imagination: having chosen to base its own narrative on experimental verification, it has had to come up with ever more inventive techniques and bolder visions. In order to explore the more hidden corners of matter and of the universe, science has had to overcome every limit and has turned the story of origins into an extraordinary journey.

In doing this it has frequently had to change the paradigms of humanity's way of thinking about things. It has done this many times throughout history, from Anaximander to Heisenberg and Einstein, and continues to do so. Science constantly advances, and it changes our way of seeing and describing the world. Whenever this happens, everything changes. Not just because of the new instruments and technologies that arise from it, but above all because of the fact that changing paradigms modifies all of our relations. When we look at the world with different eyes, our culture changes along with our art and philosophy. To understand and anticipate these changes is to have the tools to build a better human community.

For this reason art, science and philosophy are still essential disciplines, giving consistency to our being as humans. This unified vision of the world, which originates from our most distant past, is still the most suitable tool to deal with the challenges of the future.

Epilogue: The Massacre of the Assumption

Modica, 21 February 2018. The Val di Noto in Sicily is full of gems – but when you arrive in Modica, especially at night, you cannot fail to be enchanted. It's a city split in two by the spur of the Pizzo hill, dominated by the Castello dei Conti. Its houses lean against each other, covering the flanks of the mountains in which ancient cave complexes still have their entrances. Numerous baroque churches rise above imposing stairways. I was not expecting Modica to be this wonderful.

I'm here to talk about the origin of the universe at a conference tomorrow that's dedicated to the philosopher, doctor and scientist Tommaso Campailla. The city in which he was born, in 1668, has decided to commemorate his birth, borrowing the title of its celebrations from his most important work: *The Adam, or, The Created World*. Campailla, a distinguished follower of Descartes who corresponded with some of the leading lights of the period and was visited in Modica by George Berkeley, no less, wrote the philosophical poem *The Adam* as a kind of verse compendium of creation. Tomorrow, using this as a springboard, we will talk about the Bible and Genesis, creation and science. Besides myself they have invited Shalom Bahbout, the Chief Rabbi of Venice, as well as the Jesuit theologian Father Cesare Geroldi.

Tonight we are having supper together, guests of an excellent restaurant run by a Jewish family, with a menu that is strictly kosher. Sat around the table with us are representatives of the small local Jewish community that is raising funds to reopen a synagogue in the town. While we are eating, someone mentions the massacre of the Assumption, a distant episode in the history of Modica which has cast a long shadow over the life of this ancient community.

In 1474 the city had for centuries sustained a Jewish community, almost all living in Giudecca, the Jewish quarter. On the feast of the Assumption, a famous Dominican priest and fiery orator, Giovanni da Pistoia, arrived from Ragusa to preach and celebrate mass in the church of Saint Mary of Bethlehem. For some time the prevailing practice of conversion preaching had targeted Jews, who were forced to listen to such attempts to convert them. This had become commonplace, and usually occurred without incident, but that Sunday something went badly wrong. A tumult broke out in the crowd, incidents occurred, there were deaths. A crowd armed with pickaxes, knives and work-tools attacks the Jews present and bloodies the altar. Shouting 'Viva Maria! Death to the Jews!', they slaughter men, women and children; then the mob heads for the Giudecca neighbourhood and attacks every house there. Hundreds of people are murdered, all the homes are sacked, and the synagogue set on fire. A brutal hunt for Jews lasts for days. The few survivors of the hideous pogrom hide in caves or flee to other cities in search of protection. There has never since that Sunday been a place of worship for Jews in Modica, and having been forced to go through seemingly endless ordeals, including Italy's racial laws and deportation, the descendants of that small community are seeking to build a synagogue.

The next day at the conference I am scheduled to speak first, and I give an account of the birth of the universe according to science. Then it is the turn of Cesare Geroldi, a Jesuit theologian who has lived for many years in Jerusalem and has edited a new translation of the book of Genesis. Father Geroldi has an imposing physical presence; he is charming, charismatic, and a great storyteller.

His opening words that day are certainly striking: 'Professor Tonelli has given an account of the birth of the universe. What he has told us is the most precise account of what happened 13.8 million years ago, in the very remote past. I will speak instead about Genesis: a book which talks about the future.' And he explains that in order to understand the book of Genesis, we must start from the epoch and the context in which it was written.

There now seems little doubt that the book is in fact two books, written in different eras and by many different hands before being integrated in the first book of the Torah. This biblical scholar cites the many contradictions between the two versions. He underlines the differences in language and style, and the two different kinds of narrative perspective on the same events, in which not only is there no agreement as to their sequence – plants and animals created before mankind in one version, and after him in another – but even the name of the main protagonist changes: the *Elohim* of the first version turning into the unpronounceable *Yhwh* of the second.

But the most important thing comes after this, when he gives an account of the context in which the most sacred of books was written. We have to imagine ourselves in Babylon in the sixth century BC. Nebuchadnezzar II, having stormed

Jerusalem and destroyed the Temple, has deported the religious, social and intellectual elite of the Jewish people. It is the most terrible affliction, and for the ancient religion of Abraham and Moses it seems like the final hour has come. The proudest members of the chosen people, having been humiliated and stripped of their lands, now find themselves confronting the vastly superior power of their conquerors – a kind that is not just material and military. Nebuchadnezzar, king of the universe, represents a civilisation unequalled at the time. Babylon is the biggest city in the world, and is resplendent with many marvels; its scholars excel in all the disciplines, and have collected in thousands of tablets and papyruses the accumulated knowledge of millennia.

Faced with a civilisation based on the written word that the Assyrian-Babylonians developed, the Jewish elders decide to also put together for the very first time in written form the story of the origins of the Jewish people. In the most terrible, desperate circumstances they produce a text to cling to that represents their identity, their deepest roots. To this sacred book they commit their hope of overcoming the chain of misfortunes that has befallen them: by giving an account of the origins of the world they seek their future, dreaming of returning to Jerusalem and rebuilding the Temple and their glorious civilisation.

It is the same kind of response that generations of Jewish families have had to cultivate, for millennia, when faced with the hardest of tests. Holding fast to the Bible, they will be able to endure the most horrifying persecutions. This was no doubt the case for the small groups of Jews who survived the massacre of the Assumption in Modica.

And this is what gave me the idea of writing this book, and of calling it *Genesis*. To allow everyone to have the great story of origins that modern science has given us, to understand our deepest roots, and to find ideas with which to face the future.

Acknowledgements

I would like to thank all those who through arguments and discussions have provided prompts for this book: Sergio Marchionne, Father Cesare Geroldi, Rabbi Shalom Bahbout, Remo Bodei, Monsignor Gianantonio Borgonovo, Vito Mancuso, Pippo Lo Manto, Piero Boitani, Sonia Bergamasco and Lucia Tongiorgi.

I am particularly grateful to Alessia Dimitri, without whose determination this new adventure would not have begun.

Finally, a special thanks must go to Luciana, not only for her patience during the period of intense overtime that the first draft of this book required, but for the endless suggestions, the numerous discussions about art and philosophy, and the accurate reading of the manuscript that made it possible to develop and improve many parts of the text.

A NOTE ABOUT THE AUTHOR

Guido Tonelli is an Italian particle physicist who played a key role in the discovery of the Higgs boson, the so-called God particle, which earned François Englert and Peter Higgs the 2013 Nobel Prize in Physics. For his contributions to the field, Tonelli was made a commendatore of the Order of Merit of the Italian Republic in 2012 and was awarded the Enrico Fermi Prize from the Italian Physical Society and the $3 million Special Breakthrough Prize in Fundamental Physics. He is a professor of general physics at the University of Pisa and a visiting scientist at the European Organization for Nuclear Research (CERN).